AF369179

From Dust to Design: A Predictive Approach to Gas Turbine Durability

Shiva

Copyright © 2024 by Shiva

All rights reserved. No part of this book may be reproduced in any man-ner whatsoever without written permission except in the case of brief quotations embodied in critical articles and reviews.
First Printing, 2024

Table of Contents

Chapter 1. Introduction

Despite recent advancements in renewable energy sources, gas turbine engines remain the predominant source for aircraft propulsion and electrical power generation applications. The gas turbine industry has grown rapidly over the past several decades due largely to burgeoning commercial aviation markets in China and Africa as well as increased performance demands in military theatres in the Middle East. The operation of air-breathing engines in these regions has introduced unique engineering challenges for engine manufacturers due to the high concentrations of loose ground and airborne particulate in these geographical as depicted in Figure 1.1.

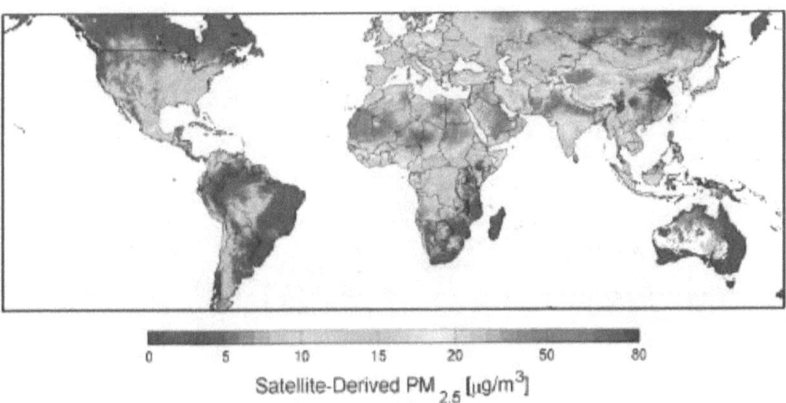

Figure 1.1: Contour of airborne particulate density for particles smaller than 2.5 micron [1].

An understanding of the general principles of gas turbine operation is needed to understand the adverse effects of particulate ingestion in these engine environments. Figure 1.2 shows a simple schematic for a typical turbofan jet engine. A fan draws ambient air through the inlet, increasing the total pressure of the air near the outer radius. This energized mass is routed through a bypass duct between the core of the engine and the nacelle and accelerates through an exit nozzle generating thrust. The remaining air closer to the fan shaft enters a compressor section in the engine core. The compressor does mechanical work on the air, raising the pressure and temperature of the gas. The air then enters a combustion chamber where it is mixed with a liquid propellant and combusts, adding thermal energy to the working fluid. The gas then expands through a turbine

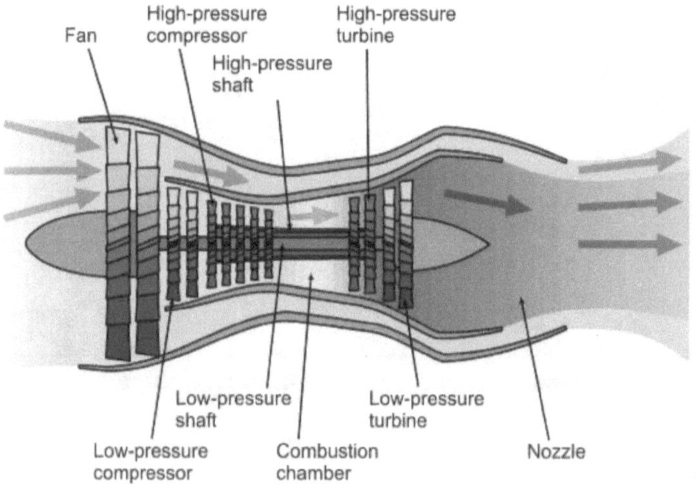

Figure 1.2: Turbofan jet engine schematic [2].

section that extracts energy from the fluid to spin one or more shafts. The shafts rotate the fan and compressor, and in the case of power generating turbines also spin a generator. In aviation engines, the hot gas in the core exhausts out of a nozzle, which increases the gas velocity and also generates thrust. The fraction of air that enters the bypass is dictated by the operational requirements of the aircraft.

Gas turbines operate on the thermodynamic Brayton cycle. A simple energy balance of this cycle reveals that thermal (fuel) efficiency can be improved by compressing the air to higher pressures and temperatures before combustion, as shown in Equation 1.1.

$$\eta_{th} = 1 - \frac{1}{PR^{\gamma-1/\gamma}} \qquad (1.1)$$

As compressor pressure ratios (PR) have pushed to higher levels, the gas temperatures in the combustor and turbine sections have far exceeded the thermal material limits of the metal hardware, which are typically cast out of high temperature nickel-alloys. Various cooling schemes are employed to reduce the metal temperatures to safe levels. The coolant air is bled from the compressor and routed to cooling passages inside the different components. Figure 1.3 depicts some of the common cooling techniques in a turbine blade. After traversing the internal cooling channels, much of the coolant impinges on the inner wall of the blade. Impinging jets provide high heat transfer coefficients and are typically utilized in regions which require enhanced cooling, such as along the leading edges of vanes and blades. After impinging and cooling the inner surface, the flow exhausts through film cooling (effusion) holes, creating a thin layer of coolant on the external surface of the vane, protecting it from the hot combustion gases. Combined impingement and effusion cooling architectures are utilized throughout the hot section. In turbine vanes and blades, additional coolant traverses through serpentine channels inside the blade where ribs and pin fin arrays are strategically placed throughout to maximize cooling effectiveness.

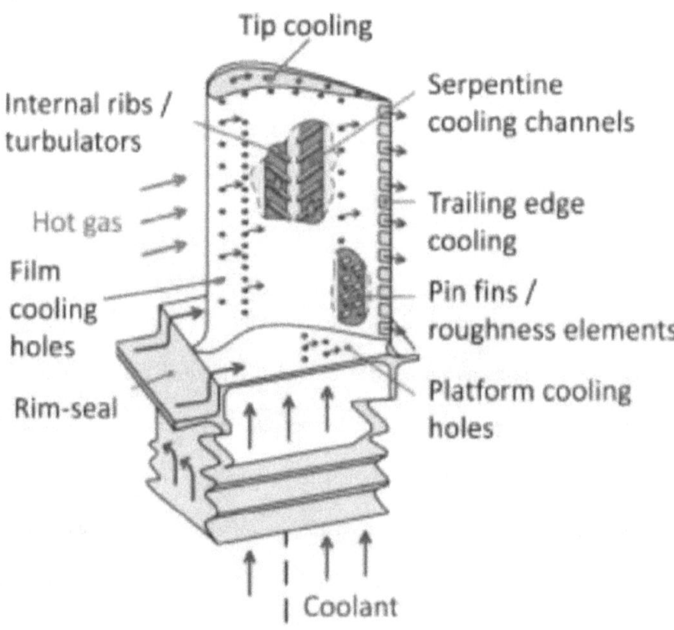

Figure 1.3: Typical turbine blade cooling schemes [3].

When particulate is ingested in an air-breathing engine it can have a range of detrimental effects depending on the sections it interacts with. Fans and compressor blades are susceptible to erosion and pitting that alter the aerodynamic surfaces and can cause inefficiencies. Erosion is the consequence of high-energy collisions between dense mineral particles and high-speed rotating components. The temperatures in the fan and compressor are well below the sintering temperatures of common particulate and allow particles to maintain their hardness. In the hot sections of the engine, impacting particles tend to stick and build deposit structures rather than eroding the original hardware. These deposition events and their associated impact are the focus of this work.

Engine deposition can be broadly split into two classifications: external and internal. External deposition refers to particles in the hot combustion gases that accumulate on the outer surfaces of the engine components, while internal deposition aptly describes agglomerations that build in the internal cooling schemes. There are certainly commonalities in the governing physics that dictate both phenomena, but these are often thought of separately because of mechanisms unique to each. External deposits are typically formed by large particles that deviate from the fluid streamlines and ballistically collide with surfaces. The turbine inlet temperatures in modern engines like the Rolls-Royce Trent 7000 can exceed 1800 K [4], which causes sintering, melting, and changes in chemistry to drive the external deposition process. Multiple studies have revealed that the onset of deposition on typical external engine hardware occurs between approximately 1200-1350 K [5][6]. Figure 1.4 shows section of a turbine vane annulus in a Boeing 747 powered by a Rolls-Royce RB-211 engine that flew through a volcanic ash cloud from the Galunggung volcanic eruption in Indonesia [7]. The deposit structures seen on the pressure side of the vanes drastically alter the aerodynamic profile of the airfoil. In moderate cases this will lead to off-design performance and

Figure 1.4: Volcanic ash deposit on a Boeing 747 turbine nozzle guide vane annulus [7].

reduce the propulsive efficiency of the engine. In severe cases like the one depicted, the throat area of the nozzle can experience a significant reduction that limits the mass flow through the engine. A rapid reduction in mass flow can lead an engine to stall or surge, resulting in a flameout and a violent yawing of the aircraft. Although the presence of deposit can provide an extra layer of thermal insulation, the blockage of film cooling holes results in a net negative thermal impact that can lead to thermal fatigue and material failure.

Internal deposition conversely is dominated by the interaction of small particles, typically below 3 μm, which tend to follow fluid streamlines more readily. This causes particles to get trapped in complex secondary flow structures in the cooling passages. The reduced temperature in the coolant path tends to mitigate issues with changes in the mechanical properties of the particulate, and the interaction is instead dictated by the adhesive forces between the particle and the surface. Deposit on the internal surfaces creates an insulative layer that prevents heat from being transferred to the coolant effectively, increasing the metal temperature. The structures alter the shapes, increase the surface roughness, and reduce the flow areas of novel cooling schemes. Each of these effects result in a reduction in coolant mass flow rates for a fixed internal to external flow PR. This work will focus specifically on aspects of the formation and impact of internal deposits, however, the physics investigated and modeling techniques employed will have relevance for external engine flows.

Chapter 2. Deposition Prediction and Modeling

Gas turbine designers and engineers have a pressing need to identify and produce new or modified components that are tolerant to operation in dust laden environments. Additionally, engine manufacturers need to understand the timescales of deposition induced part degradation so they can forecast maintenance schedules to service and repair impacted hardware. While the majority of these decisions are currently made through experimental investigations, ultimately the goal is to obtain a predictive modeling capability that will reduce the number of resources required to develop suitable designs.

Predicting deposition first requires the ability to both model the various flow-fields as well as the interaction of the flow and the particulate. Computational Fluid Dynamics (CFD) solvers are a popular and reliable tool to predict both physics. CFD programs utilize a mesh that reconstructs the geometrical domain of interest with a series of smaller cells. The Navier-Stokes equations are then discretized and solved on the grid given proper boundary conditions, yielding a solution for the flow pressure, temperature, density, and velocity field. These methods describe the flow solution in an Eulerian frame.

Once the flow is solved, particles can be tracked through the flow using a Lagrangian tracking scheme. The motion of the particle is solved using a Newtonian formulation as shown in Equation 2.1. The right side of the equation represents the various forces causing the particle acceleration. F_D is the fluid drag force, the middle term includes the gravity and buoyancy forces, and the final term represents other forces that may affect the particle motion (e.g., thermophoresis). This equation is broken down into three equations, one for each component of the particle velocity vector (u_p).

$$m_p \frac{du_p}{dt} = F_D + \frac{g(\rho_p - \rho)}{\rho_p} + F \qquad (2.1)$$

Tracking the particles through the fluid domain is a mature practice, and with the correct inclusion of all of the forces will result in realistic trajectories. Particle collision modeling conversely is a relatively new area that is experiencing a thrust among the scientific research community. The focus on the development of "bounce-stick" models has been spurred on by the gas turbine industry which desires the ability to accurately predict where particles stick or how they rebound back into the fluid.

Deposition models are divided into two types: stochastic and deterministic. The former generally rely on statistical data to identify thresholds for sticking. These are appealing because they can provide very good matches to experimental data and are computationally efficient, however they can suffer from lack of generality and require a large and diverse experimental deposition database. Although deposition models all require some degree of probabilistic treatment due to the irregular shapes and tendency for particles to spin, stochastic models often have apparent physical contradictions in the sticking prediction. Some notable stochastic models include the EBFOG model by Casari et al. (2017) [8], the Critical Viscosity model by Sreedharen and Tafti

(2011) [9], and the Rolls-Royce Sticking Model by Barker et al. (2017) [10]. Deterministic models on the other hand strive to capture the individual particle mechanics through a physics-based approach. While these have a higher degree of fidelity, they are more computationally expensive and require extensive knowledge of the both the relevant physical processes and the material properties and constants. The Critical Velocity model by Brach and Dunn (1992) [11] and the OSU Deposition Model by Bons et al. (2016) [12] are two popular deterministic models.

The predictive capability of these models relies on a complete understanding of the variable space for both the aerodynamic trajectories and impact mechanics. Stochastic models often identify a few key variables and assign thresholds, but physics-based models require a more comprehensive approach as they strive for generality. Numerous studies have identified temperature as a primary variable affecting deposition rates. Smith et al. [13] varied the gas temperature in an accelerated nozzle guide vane test facility and found that deposition of coal ash particles on the vane external surfaces increased substantially with temperature. Whitaker et al. [14] maintained a constant coolant temperature (866 K) in a nozzle guide vane leading edge simulator and varied the metal temperature with an electric kiln and discovered a linear relationship between coolant mass flow blockage and surface temperature. Bowen et al. [15] developed deposit cones through an oversized canonical impingement jet and demonstrated that deposit can accumulate on surfaces at temperatures as low as room temperature.

These different test campaigns distinguished the role of the flow and the surface temperature, but it is still unclear what the role of the particle temperature is. Test facilities with control over the temperature of the fluid phase are designed to assure that the injected particulate accommodates to the flow temperature before interacting with the test section. Bonilla et al. [16] performed deposition testing on the external surfaces of two CFM56-5B doublets at a constant freestream temperature and Mach number, but varying the film cooling back flow margin (BFM) and temperature. The authors found that they could reduce the buildup on the surface by reducing the coolant temperature. They emphasized that the particle temperature upon impact is a function on the amount of time it spends traversing the film cooling layer in the near wall region. Figure 2.1 shows a general temperature distribution in the near wall region from a Large Eddy Simulation (LES) film cooling study by Baek and Yavuzkurt [17]. A particle in the hot main flow that enters this thermal boundary layer will cool to some intermediate value between the freestream and wall temperatures before impact. The time dependent cooling of the particle is usually modeled with a 1-D lumped model as given by Equation 2.2. The rate that a particle accommodates to the local temperature in the thermal boundary layer is a function of the rate of heat transfer from the particle

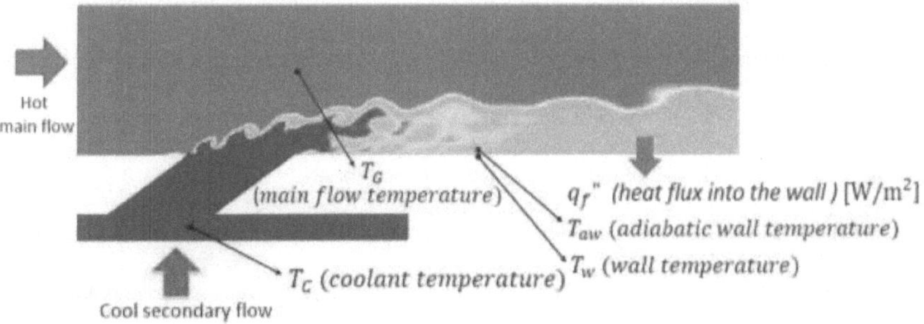

Figure 2.1: Temperature contours from instantaneous Large Eddy Simulation (LES) showing the film cooling distribution [17].

$$m_p\, c_p \frac{dT_p}{dt} = hA_p\left(T_\infty - T_p\right) \tag{2.2}$$

$$\tau_t = \frac{m_p\, c_p}{hA_p} \tag{2.3}$$

to the coolant, and the thermal capacitance of the particle ($m_p\, c_p$). The thermal time constant (τ_t), or time that it takes for the temperature to reach 63.2% of the initial (hot) and final (cool) temperature is given by Equation 2.3. The particle diameter and density are clearly important, as the particle mass will scale roughly by the cube of the diameter assuming spherical particles.

The particle and surface temperatures at impact in engine flows can vary and both appear to play a crucial role in determining the likelihood of sticking. In external flows where the particle temperatures often exceed the melting temperature, it's logical that the particle temperature dominates since the changes in particle properties and shape can cause liquid physics like wetting and splashing that are rather independent of the surface temperature in general. For impacts in internal cooling channels at temperatures below sintering points, the adhesion binding the particle to the substrate is due to numerous forces. An incomplete list includes Van der Waals forces, electrostatic forces, mechanical interlocking, and magnetic forces.

Van der Waals forces are caused by temporary dipole interactions between objects in contact due to the constant shifting of electrons in molecules at the particle surface. These are thought to be the main force involved in low temperature deposition, but the scientific literature is sparse in terms of temperature dependencies for Van der Waals forces. The few studies available for materials not relevant to minerals in common dust show an inverse proportionality between Van

Der Waals forces and temperature. The complexity involved with quantifying the numerous short-range molecular forces that contribute to the overall adhesion has led scientists to make simplifications. The surface free energy is a material constant often used to estimate adhesive bonds in materials and can be thought of as the work required to cut or separate a bond. Young and Duprè [18] developed a simple formula to determine the work of adhesion (W_a) between a liquid and solid in contact for wetting applications as shown by Equation 2.4. Although particles in coolant channels are not a liquid phase, the particles still undergo some elastic and plastic deformation during the collision with the wall and thus the wetting deformation analogies maintain some relevance. In Equation 2.4, the work is equal to the sum of the individual surface energies of the objects in contact (γ_1 and γ_2) minus an interfacial resistance (γ_{12}) resisting the adhesion.

$$W_a = \gamma_1 + \gamma_2 - \gamma_{12} \tag{2.4}$$

Particle diameter is another variable that has been extensively explored by deposition researchers. Wolff et al. [19] conducted experiments using different size distributions of Arizona Road Dust (ARD) in an effusion cooling array representative of a combustor cooling liner. Their results confirmed the results of many other studies that internal deposition favors small particles. Additionally, they found that for internal deposits the threshold for deposition appears to be below 3 µm, a finding originally published by Whitaker et al. [20]. This study also elucidated the role of larger particles (> 5 µm) which can erode and remove existing internal deposition structures.

Particle diameter is fundamental to deposition because it affects the aerodynamic trajectories of particles as well as the mechanical impact physics. The most common dimensionless parameter used in deposition studies is the Stokes number, which measures the ratio of the particle inertia to the particle drag using a Stokesian Drag assumption as shown by Equation 2.5. The Stokes number will be explored more in detail in a latter section, but it is generally thought that a particle with a Stokes number smaller than 1 is drag dominated and follows flow streamlines, while particles with Stokes numbers greater than 1 are inertia dominated. The former are less likely to impact surfaces while the latter are more ballistic and more prone to hitting surfaces. The Stokes number scales with the square of the particle diameter, so larger particle size distributions are far more likely to impact. This does not however mean that large particles are more likely to stick as the conclusions of previous studies determined.

$$St_k = \frac{\rho_p d_p{}^2 \left(U_f - U_p\right)}{18\mu_f} \tag{2.5}$$

The sticking probability is determined by the mechanical interaction of the particle and the surface. This boils down to an energy balance between the incoming particle kinetic energy and the work of adhesion acting to resist the rebound energy of the particle. The kinetic energy scales with the mass and thus the cube of the diameter, while the adhesion scales with the contact area of the particle and the surface, which varies approximately with the square of the diameter. This biases sticking to small particles. The capture efficiency (η_c) is a common metric in deposition measuring the percentage of total particles delivered that eventually deposit (Equation 2.6). It is equal to the product of the impact efficiency (η_{imp}) and the sticking efficiency (η_{imp}). Deposition events are dependent on too many specific variables to prescribe a definite relationship between the capture efficiency and particle diameter, however, in most cases relevant to engine cooling the trends are non-monotonic. While small particles may have a high sticking efficiency, their impact efficiency is lower. Conversely, large particles impact readily but are more unlikely to stick. The net result is a maximum capture efficiency at some intermediate size where particles are sufficiently large to impact, but sufficiently small to deposit.

$$\eta_c = \frac{\#\ of\ sticks}{\#\ of\ particles} = \eta_{stick} * \eta_{imp} \tag{2.6}$$

$$\eta_{imp} = \frac{\#\ of\ impacts}{\#\ of\ particles} \tag{2.7}$$

$$\eta_{stick} = \frac{\#\ of\ sticks}{\#\ of\ impacts} \tag{2.8}$$

Temperature and particle size are variables that emerge in nearly every deposition study and as such deserve specific discussion, but the full list of parameters that are known to influence deposition is practically endless. Table 1 provides an incomplete list of variables that affect both the trajectory and impact physics respectively. The extensiveness of the parameter space limits the applicability of empirical models derived for a specific set of boundary conditions in most circumstances. Many of the variables interact with both the trajectory and mechanical physics and create competing effects in terms of predicting deposition rates, as was demonstrated with particle diameter. Attempts by researchers to replicate deposition results through scaling and similarity laws as is common in other engineering disciplines is futile. Suman et al. [21] attempted to identify the most pertinent physics and non-dimensionalize the data from over 70 external deposition tests from different researchers and found that mapping the results was unsuccessful due to small differences in certain parameters. This highlights the need for a universal model that can simplify the physics to the most deterministic processes and model those with a high degree of fidelity.

The role of fluid pressure on deposition was notably missing from the literature until 2016 when Sacco et al. [22] published the first results in a high-pressure deposition facility. The first part of this work will focus on understanding the physics associated with deposition in high-pressure engine environments through an experimental and computational campaign. The results will reveal that the effect of pressure is perhaps as influential as temperature and should garner the same level of detail and consideration by the deposition community.

Table 2.1: Table of variables affecting particle trajectories and impact mechanics.

Trajectory	Impact
Particle diameter	Particle diameter
Particle density	Particle density
Particle sphericity	Particle sphericity
Fluid viscosity	Particle temperature
Fluid density	Particle velocity
Fluid temperature	Surface temperature
Fluid velocity	Fluid viscosity
Particle rotation	Velocity gradients
Reference frame rotation	Turbulent kinetic energy
Particle loading	Particle mechanical properties
Thermal gradients (thermophoresis)	Surface mechanical properties
Velocity gradients (Saffman lift)	Particle rotation
Turbulent kinetic energy (turbophoresis)	Reference frame rotation
Fluid/particle/surface charge	Fluid/particle/surface charge

Chapter 3. Deposition Studies at Elevated Pressures

The pressure of the fluid carrying particulate in an engine varies depending on the section of the engine being studied as well as the altitude and operating characteristics of the particular engine. As discussed previously, the compressor section in a turbomachine performs shaft work on the fluid to increase the pressure. This allows for more efficient combustion and higher thermal efficiency, and modern engines are pushing the limits of the overall pressure ratio (OPR) that can be reached in pursuit of higher efficiency and increased performance. Figure 3.1 shows how the OPR of turbine engines have increased from the 1960's to the mid 1990's. The most current engines, like the General Electric GE9x, are advertising an unprecedented OPR of 60:1. At takeoff, this translates to combustor and turbine inlet pressures of roughly 48.6 atmospheres (atm), while at cruise conditions this drops to approximately 14 atm. Deposition events can occur at takeoff when the fan ingests loose-ground particulate as well as at higher altitudes where the aircraft can fly through volcanic ash clouds or other particulate laden air.

The pressure ranges involved in actual engine deposition are quite considerable, but the effect of pressure on accumulation rates has been understudied generally by the deposition research community. This can be contributed primarily to the limitations of deposition experimental testing. The most realistic testing results will always be achieved via full-engine testing, where the hardware is authentic and all of the boundary conditions relevant to deposition can be matched. The cost to run a full-engine test however is exorbitant and requires significant man-hours to set up and run safely and correctly. Two additional disadvantages are that it is difficult to obtain flow

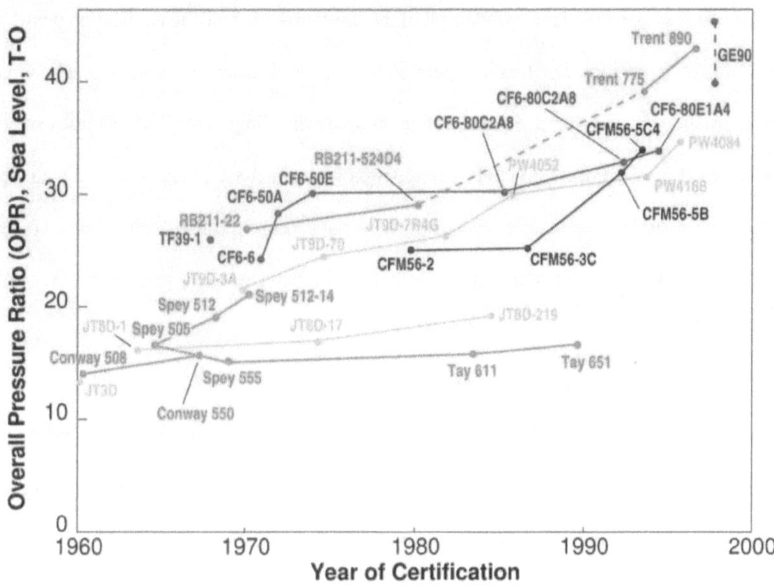

Figure 3.1: Gas turbine engine overall pressure ratio (OPR) trends [23].

and surface measurements, and there is little to no control over particular independent variables that might be of interest.

Scaled down laboratory facilities are a more appealing option because they satisfy many of the downfalls of full engine testing. Many of the experimental rigs designed to mimic engine conditions can be operated cheaply and safely and require less manpower. The facilities are designed with measurement access in mind and are simplified so that independent variables can be changed without affecting other parameters. The question that emerges with any experimental facility is whether it appropriately models all of the physics in the real-life application. Until 2017, every academic deposition test facility operated with air at or near atmospheric levels. This is a common practice because it reduces the design complexities, limits cost, and increases safety for the operators. These facilities are vital to amassing a large database of experimental data that are used to inform the bounce-stick models, but there have been questions regarding whether the mismatched absolute pressures involved result in deposition patterns that do not match field serviced engine hardware.

In 2017, Sacco et al. developed the first academic level deposition facility capable of achieving absolute flow pressures up to 17 atm. While this was still a far cry from the 40 atm typical at take-off in modern engines, it allowed for the effect of increasing absolute pressure to be elucidated. The facility housed an inline flow heater, dust delivery mechanism, and test section in respective pressure-vessels. Compressed air at 136 atm was regulated down to the necessary pressure level upstream to set the desired pressure in the vessel housing the test article. The test article studied was an impingement hole array containing 8 columns and 7 rows of 0.65 mm diameter holes. The impingement plate separated from a solid stainless-steel target plate to give a gap-width to hole diameter ratio (z/d) of 2.2. These dimensions are typical in an engine impingement cooling scheme. A schematic of the test assembly is illustrated in Figure 3.2. Due to the spacers blocking the flow on the top and bottom edges of the plate, the flow was forced to turn out of the sides and developed a crossflow in each direction as shown in the bottom schematic of Figure 3.2.

The authors delivered nominally 2 grams of 0-10 μm ARD over a 2-minute period. They carried out this procedure at three flow temperatures (293 K, 505 K, 728 K) and two fixed PR across the impingement holes (PR = 1.015 and 1.03), and at 6 cavity pressures ranging from 1 atm to 14.6 atm. This allowed for the trends in pressure to be evaluated alongside two other significant variables known to impact particle sticking probability.

Figure 3.3 is a plot from Sacco et al. of the capture efficiency versus pressure for each of the three temperatures at a PR of 1.015. The findings indicated that at elevated flow temperatures, the deposition on the target surface decreased non-linearly with pressure and asymptotically approached a capture efficiency of zero. The overwhelming takeaway from the work was that deposition tests run at atmospheric pressures can overestimate the magnitude of deposition by more than an order of magnitude. Any deposition experiment run with the goal of understanding how particles will accumulate in a high-pressure engine environment then must match the absolute flow pressure as well as other variables like temperature and pressure ratio.

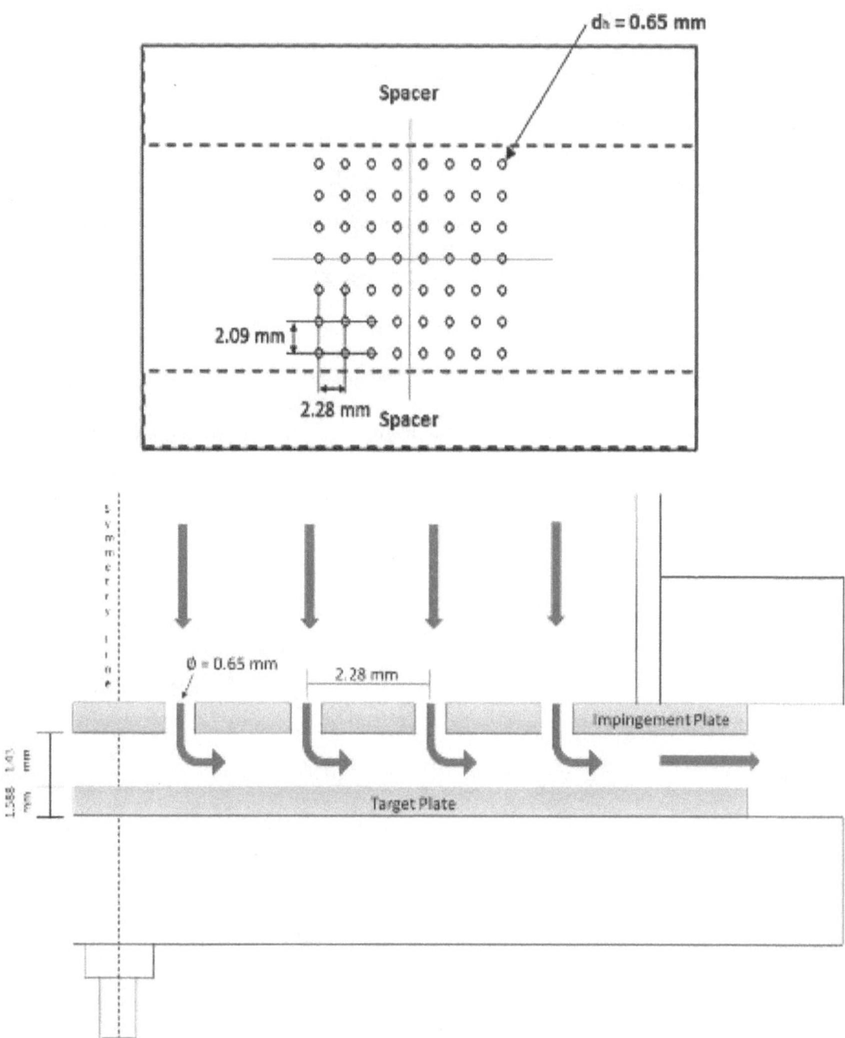

Figure 3.2: Sacco et al. impingement array geometry schematic [22].

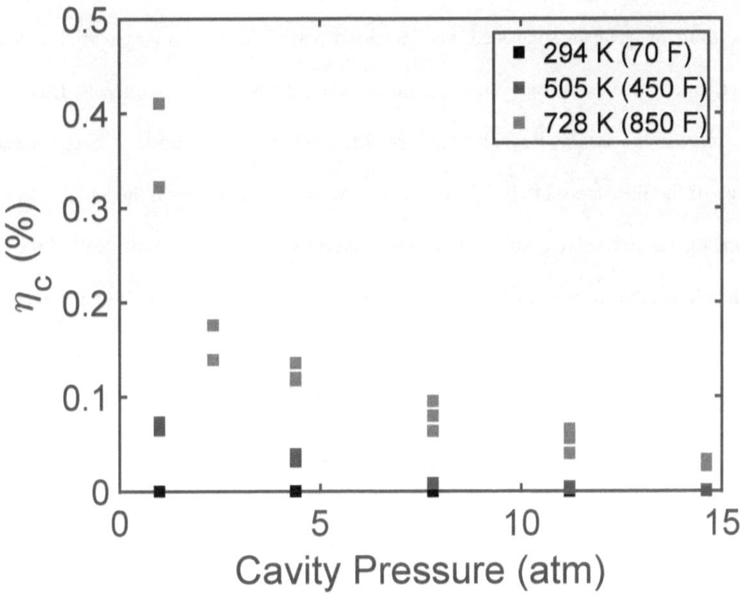

Figure 3.3: Capture efficiency (η_c) vs. cavity pressure and temperature at PR = 1.015 in Sacco et al. impingement array experiments [22].

While the effects that pressure can have on deposit formation in a common cooling geometry were emphasized by the results of Sacco et al., the physical mechanisms responsible were still unclear. The authors suggested three mechanisms that were likely candidates for explaining the reduction in accumulation at higher pressures. All three mechanisms emerge as a consequence of an increasing Reynolds numbers with fluid density due to increasing pressure. The first two mechanisms are increased shear removal forces and elevated discharge coefficient. These two phenomena are a result of an increasing flow hole Reynolds number (Re$_d$), defined in Equation 3.1. The third mechanism suggested was increased particle drag allowing particles to follow streamlines more readily and reducing the number and the nature of surface impacts. This is caused by an increase in the particle Reynolds number (Re$_p$), defined by Equation 3.2. Each parameter is a measure of the flow inertia to the flow viscosity, which is a function of the length scale of the object and the relative velocity of the flow to the object of interest. The length scale in the case of the cooling hole is the hole diameter (d), while for the particle it is the particle diameter (d_p).

$$Re_d = \frac{\rho_f U d}{\mu_f} \tag{3.1}$$

$$Re_p = \frac{\rho_f (U_f - U_p) d_p}{\mu_f} \tag{3.2}$$

The broader objective of this research is to improve deposition modeling techniques and practices. The goal of the first part of this work is to determine the importance of each of these mechanisms through an experimental and computational analysis of deposition effects as a function of pressure in an effusion cooling geometry. This is critical to informing what physics can be ignored, which may be simplified, and which need careful treatment in the modeling process.

Chapter 4. High-Pressure Experimental Setup

The experiments are performed in the High-Pressure Deposition Facility (HPDF) at The Ohio State University Aerospace Research Center. This facility is an upgraded version of the one developed and use by Sacco et al. in 2017. The primary upgrades include pressure vessels capable of withstanding pressures up to 23.8 atm, and an inline flow heater with 6 times more power, allowing for hotter flow temperatures to be achieved at higher mass-flow rates.

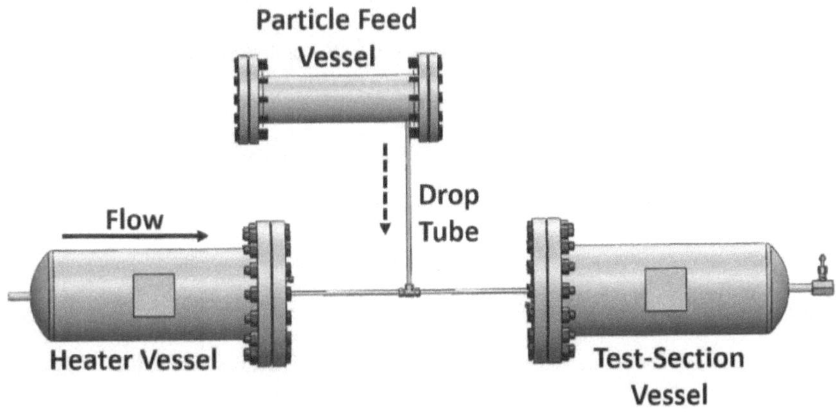

Figure 4.1: HPDF facility schematic.

A facility schematic is provided in Figure 4.1. Compressed air at approximately 136 atm is regulated to a safe value depending on the proof pressure of the mass flow controller used for the test. The mass flow controller that is used depends on the mass flow requirements for the test, which scales proportionally with the particular cavity pressure. One mass flow controller is utilized for the lower pressure/mass flow tests and has an upper limit of 0.005 kg/s with accuracy of ± (0.8% of reading + 0.0001 kg/s). This controller is necessary for decreased measurement uncertainty at these lower flow rates. The second controller is rated for high pressures and has an upper limit of 0.0592 kg/s with accuracy of ± (0.8% of reading + 0.0012 kg/s). The maximum uncertainties in mass flow rates for these respective controllers occur at their lowest tested pressures and are ± 1.04% and ± 6% respectively for the conditions run for this campaign. The air passes through the mass flow controller and is introduced into the first pressure vessel titled the "heater vessel" in Figure 4.1. This vessel houses a 36-kW inline flow heater capable of reaching temperatures at the exit of 1144 K. The heated gas exits the first vessel and is plumbed through 0.3 m of 15.88 mm diameter stainless steel tubing.

Above the main flow tubing, a "particle-feed vessel" rests on a unistrut structure. Inside the vessel is a conveyor belt that is controlled remotely. Particles on the belt fall through a funnel and down through the vertical tubing section before entraining in-the hot gas. The hot particulate laden gas then enters a final 0.3 m of tubing welded through the flange of the "test-section vessel" before it interacts with the test article and exhausts through the effusion holes. The exhausting gas fills and pressurizes the vessel before it exits a needle valve. The vessel operates as a choked nozzle and thus small adjustments can be made in the needle valve area to make fine adjustments to the cavity pressure for a given mass flow rate.

Two k-type thermocouples with standard limits of error of ± 5.5 K are used to obtain temperature measurements for each test. One is placed 25.4 mm upstream of the effusion plate and is centered in the test-section channel to measure the flow temperature. The other is spot-welded in the center of the external surface of the plate between two columns of holes to get a representative surface temperature measurement for every test. The test-section cavity pressure is measured using a 0-68 atm pressure transducer with uncertainty of ± 0.17 atm. A second 0-3.04 atm transducer is used for test pressures below 3 atm for better accuracy of ± 0.005 atm. With cavity pressure measured, the difference between the cavity pressure and pressure upstream of the test plate is needed to obtain the pressure ratio across the plate. This differential pressure is measured with a differential transducer which measures the difference between the cavity pressure and the pressure measured 38.1 mm upstream of the test article. Each test is run at a constant PR of 1.03 in this study, and the maximum uncertainty in the PR is ± 0.0003.

Chapter 5. High-Pressure Experimental Procedure

A single effusion plate is re-used for every test. After cleaning the plate in an ultrasonic bath to remove dust stuck inside the effusion holes, the plate is weighed and recorded on a scale with accuracy of $\pm$ 0.0003 g. The plate is then fastened to the diffuser section with screws and a high-temperature gasket that compresses to approximately 0.5 mm. The same scale is then used to measure 6 grams of the 0-10 μm or 0-3.5 μm ARD, and the dust is baked for one hour at 477 K in an industrial oven to remove moisture and minimize clumping of particles. After removing the dust from the oven, it must be reweighed to account for water mass lost through vaporization during the bake. The dust is then evenly loaded onto a 0.3 m long conveyor belt in the top test vessel. A funnel is secured in front of the lip of the belt and onto the drop tube, and the particle feed vessel flange is bolted on and sealed with a gasket. The test-section vessel is then sealed, and the test begins by ramping the mass flow slowly to the desired value. The inline heater is activated and set to the appropriate value, and some minor adjustments in the mass flow, heater output level, and needle valve opening are made until the desired flow temperature, PR, and cavity pressure are achieved at a steady state condition. The conveyor belt is then turned on and run at a constant delivery speed to deliver particles to the flow and the effusion plate.

A primary metric used to characterize the level and effect of deposition on the effusion cooling performance is the mass flow reduction, or blockage of the plate. In an engine, the PR across the coolant holes is fixed even as the holes are blocked. This results in a reduction in the mass flow rate through the holes that is detrimental to the cooling effectiveness. In this setup, the mass flow is dictated by the controller. For a constant mass flow rate, a reduction in the flow area causes the PR across the holes to increase to force the fixed amount of mass through the smaller area at an elevated flow velocity. To mimic the engine operation, the mass flow setpoint on the controller is reduced to maintain the PR at the starting value of 1.03. While this is a typical operating procedure in similar low-pressure deposition rigs, there is one additional step necessary in high-pressure operation. As stated, the coolant exhausting through the effusion holes and into the vessel sets the cavity pressure. The needle valve ensures choked operation of the vessel, so a reduction in the system mass flow causes the cavity pressure to drop. This then causes the PR to increase again and the mass flow level to be reduced in turn, resulting in a positive feedback loop. The cavity pressure is a variable that should be fixed throughout the test, so the easiest method to solve this feedback issue is to inject a separate line of high-pressure mass directly into the vessel, bypassing the effusion plate altogether. In theory, the added mass through this secondary line should equal the decremented amount at the current point in the test due to blockage, to ensure the total mass into the vessel remains at the original starting value.

The test continues until the mass flow is reduced by 25% or all of the 6 grams of dust have been delivered, whichever occurs first. The maximum duration of the test is 45 minutes. This delivery rate corresponds to a loading concentration of 0.14 parts per million by volume, which is well below the threshold for a dilute suspension [20]. Having a dilute suspension of particles is necessary to ensure that the particle trajectories are one-way coupled with the flow, and particle to particle interactions are negligible. At this point the belt is stopped, and the system mass flow and heater are slowly ramped down.

Once the system has cooled, the test plate is removed and weighed. Photos and microscope images of the deposit structures are then taken. Any dust remaining on the belt is collected and weighed, and the dust stuck inside the tubing is blown into a 3-layer filter, which is weighed before and after the blowout to measure the mass stuck in the tubing. This total leftover mass was not delivered to the test section and must be subtracted from the 6 grams to accurately measure how much dust was required to interact with the plate in order to reach the 25% mass flow reduction. This leftover dust is a function of the operating pressure, so performing a careful and thorough dust recovery routine is paramount to interpreting the results.

As previously mentioned, each test is run at the same PR of 1.03 and at an upstream flow temperature of 950 K, both of which are typical values in an effusion cooling geometry in an engine. These conditions and the plate are the same as those studied by Wolff et al. at atmospheric pressure in a kiln facility. A total of seven absolute flow pressures are studied, ranging from 1 to 15.77 atm. Table 5.1 lists the pressures run, referencing the pressure measured upstream of the effusion plate. The total mass flow for each test is also documented, as well as the temperature measured on the external surface of the plate.

It is important to note that these experiments lack a realistic thermal boundary condition for the plate backside. In an engine the hot combustion gas provides a convective heat flux on the outer surface of the plate. The coolant works to reduce the temperature on the plate surface by extracting some of that heat and transferring it to the coolant. This results in a plate temperature that is lower than the hot combustion gas, but higher than the coolant temperature. The objective of engineers working on the coolant sections is to bring the plate surface temperature as close to the coolant temperature as possible using the least amount of coolant air. This performance is often measured using a parameter called the overall cooling effectiveness ($\emptyset$), shown in Equation 5.1

where T_∞ is the freestream (hot) temperature, T_w is the wall temperature, and T_c is the coolant temperature [24]. The theoretical maximum for Ø is unity which represents an ideal cooling scheme, but typical values for cooled turbine airfoils are closer to 0.6.

Table 5.1: Test matrix of pressures with corresponding mass flow and external surface temperatures.

Upstream Pressure (atm)	Mass Flow Rate (kg/s)	Plate Temperature (K)
1	0.0022	793
2.07	0.00477	842
5.86	0.014	893
7.25	0.017	897
8.76	0.021	898
12.26	0.03	902
15.77	0.038	904

$$\text{Ø} = \frac{T_\infty - T_w}{T_\infty - T_c} \tag{5.1}$$

In this rig, there is no backside heating mode to provide the energy necessary to heat the plate to engine relevant temperatures. Atmospheric facilities provide this external heat flux by placing the test article in a kiln, exposing the external surface to a torch, or by utilizing actual combustion to provide a realistic hot gas path. At high-pressure, all of these options are impractical due to lack of access to the test article inside the vessel as well as safety concerns involved with combining high-pressure and combusting flow physics. These constraints result in an external plate temperature that is cooler than the "coolant" temperature. In this work the flow will still be referred to as coolant flow because it provides that function in an engine, but in this facility the coolant is actually heating the plate. As pressure increases, the external plate temperature increases as shown in Table 5.1 due to the higher flow density in the facility and resulting increase in heat transfer coefficients. The importance of the plate temperature in deposition was previously mentioned, and it is worth considering what effect a non-constant plate temperature has on the results found in this study. The latter experimental sections will explain how this increase in surface temperature works to strengthen the conclusions that will be made regarding the effects of pressure on deposition.

Chapter 6. Test-Section Description and Dust Characterization

The test section consists of an Inconel diffuser with a flange and bolt pattern matching that of the effusion test plate that attaches to it. A model of the diffuser is provided in the top image of Figure 6.1. The inlet of the diffuser is a tube section matching the tubing welded to the flange of the pressure vessel, and the two tubing sections are connected with a stainless-steel compression fitting. The diffuser is designed to gradually transition the flow from that circular tubing to a larger rectangular channel with height spanning the vertical dimension of the effusion plate columns. Once the plate is reached, the width of the diffuser contracts. The mass flow rate through the diffuser decreases with each column of ejecting holes, which would cause the crossflow velocity through the diffuser to decrease if the area were held constant. The contraction was designed to keep the crossflow velocity constant. The flow temperature and differential pressure measurements are made in ports in this diffuser section upstream of the plate.

The effusion plate and the hole pattern are shown in the bottom left of Figure 6.1. The plate resembles a combustor section cooling liner. The plate is made of a high-temperature nickel alloy and consists of 357 total holes. The array has an alternating pattern of 16 columns of 12 holes and 15 columns of 11 holes. The column spacing is 4.06 mm, and the rows within a given column are spaced by 2.03 mm. Each hole has a diameter of 0.56 mm and a length to diameter ratio (l/d) of 4.54.

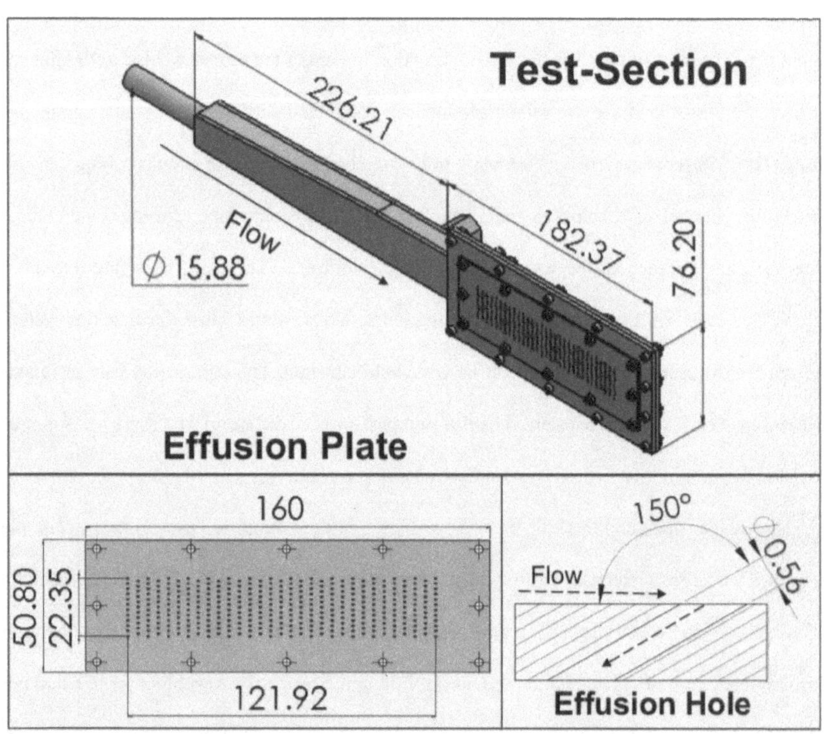

Figure 6.1: Test-section model and schematic including diffuser section (top) and effusion plate (bottom).

A recent study conducted by Varney et al. [25] did a comprehensive analysis of how the various geometric parameters in similar effusion cooling plates affect the blockage due to deposition. One of their key findings was that the hole inclination angle relative to the incoming flow direction is a key parameter. When the flow entering the coolant hole was forced to make a steep 150° turn, they found that the blockage per gram (BPG), or rate of blockage, was roughly an order of magnitude larger than when the flow made a moderate 30° turn. Since the objective in this study is to visualize and analyze how the blockage in effusion holes is affected by increasing the flow pressure, the 150° turning case was selected to maximize the amount of blockage that occurs. The bottom right image in Figure 6.1 can be referenced to visualize the steep turn that the incoming coolant nominally navigates in each effusion hole in the array.

ARD in the 0-10 μm and 0-3.5 μm range is used in this series of tests. ARD is a common test dust used in deposition experiments, as it has a mineral content and behavior that is similar to dusts encountered in other regions of the world. While any experiment or simulation performed with the intention of replicating the engine operation should take great care to characterize the particulate of the local geography in which the engine operates, fundamental testing for the purposes of understanding general physics is better performed with common tests dusts. This is due to the fact that the dust size and constituent distributions can be better controlled and measured, and there is sufficient documentation of many of the composite material properties. ARD has been used since the 1940's as a standardized dust for air filtration testing [26].

The ARD is prepared and provided by Power Technology Inc. (PTI) The chemical composition of Arizona Road Dust as provided by PTI is documented in Table 6.1. Up to 90% of the particulate mass is comprised of silicon dioxide, which occurs as quartz in nature and is the primary constituent in most sands, and alumina. Understanding the mineral composition is important to understanding the impact mechanics between the particles and the surface and between particle-to-particle interactions. Rather than modeling the various constituents and their particular properties, which would be computationally burdensome, researchers often use a law of mixtures approach to derive composite material properties. This is useful because it gives an effective value for the dust and treats it as a homogenous mixture, which is much simpler for modeling purposes. Whitaker et al. [27] documented several relevant material properties for ARD as a function of temperature using a law of mixtures approach, and those values are utilized in the latter modeling sections.

Table 6.1: Chemical composition of Arizona Road Dust (ARD) by mass percentage.

Chemical	Percentage by Mass
SiO_2	69-77
Al_2O_3	8-14
Fe_2O_3	4-7
MgO	1-2
CaO	2.5-5.5
Na_2O	1-4
K_2O	2-5
TiO_2	0-1

The cumulative mass distributions for each batch of dust studied (0-10 and 0-3.5 μm) as provided by PTI are plotted in Figure 6.2. These distributions are measured by PTI using a Coulter counter, which suspends the particles in an electrolyte solution. The particles pass through an aperture with a current attached to it, causing a measurable change in voltage which is related to the volume of the particle.

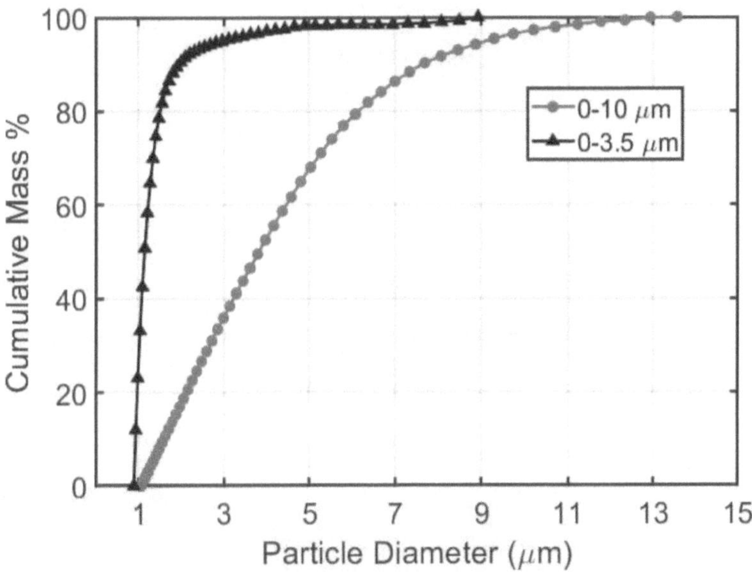

Figure 6.2: Cumulative mass distributions vs particle diameter for 0-10 and 0-3.5 μm ARD.

Chapter 7. Experimental Results

Figures 7.1 and 7.2 show the experimental mass flow reduction (MFR) versus cumulative grams delivered throughout the test for both the 0-3.5 and 0-10 μm batches respectively. Mass flow reduction (MFR) is defined in Equation 7.1 as a percentage reduction (% MFR).

$$\% \, MFR = 100 * \left(1 - \frac{\dot{m}_f}{\dot{m}_o} \right) \tag{7.1}$$

Observing the trends of Figure 7.1 first for 0-3.5 μm ARD, it is evident that every pressure tested reached 25% MFR, and did so rather rapidly, requiring less than 1 gram of dust to reach that blockage level. The trend with pressure is also clear. As pressure increases, the effusion holes block less rapidly, taking more dust to reach the 25% MFR target than at lower pressures. This monotonic trend shows that for this batch of smaller diameter dust, significant blockage occurs regardless of the pressure due to the tendency for small particles to stick aggressively upon impact, but the rate that the deposit obstructs the holes is significantly affected by pressure.

The 0-10 μm results in Figure 7.2 show an even stronger dependence on absolute flow pressure. The 25% MFR is reached only at the two lowest pressures tested (1 and 2.07 atm). At 5.86 atm and higher, the MFR asymptotes at some point, indicating that a steady condition is reached in the buildup process of the structures that block the cooling holes. This does not imply that the dust injected after the MFR asymptotes has no interaction and cannot stick, however. Dust can still stick in and around the holes, however it does so in a manner that does not contribute to obstructing the flow. Particles that do stick in areas that would otherwise create blockage might also reach an equilibrium where the mass that sticks is simultaneously eroded by large particles that impact with high kinetic energy and remove existing deposits. The 0-10 μm final MFR level and the rate of blockage also decrease monotonically with increasing pressure. The trend also appears to be non-linear, with more significant reduction in blockage occurring with increasing pressure from 1 to 5.86 atm. This non-linearity was also seen in the pressure study by Sacco et al. [22], which is logical since both tests used the exact same distribution of 0-10 μm ARD. The Sacco et al. study, however, showed a non-linear reduction in capture efficiency rather than MFR, but the MFR in these experiments is directly related to the capture efficiency.

The x-axis of both figures reveals that all of the dust is delivered for 0-10 μm at higher pressures, which occurs because the 25% MFR target was never achieved and the entirety of the 6 grams was injected. Conversely for the smaller 0-3 μm ARD, the maximum amount of dust delivered at 15.77 atm is approximately 0.55 g. Not only is a pressure trend discovered, but a significant difference is uncovered between the two size distributions of dust. These differences and the underlying physics that cause them will be vetted in a latter section. The 0-10 μm results are adequate however for investigation of the first two physical mechanisms that were suggested by Sacco et al.: shear removal and discharge coefficient effects.

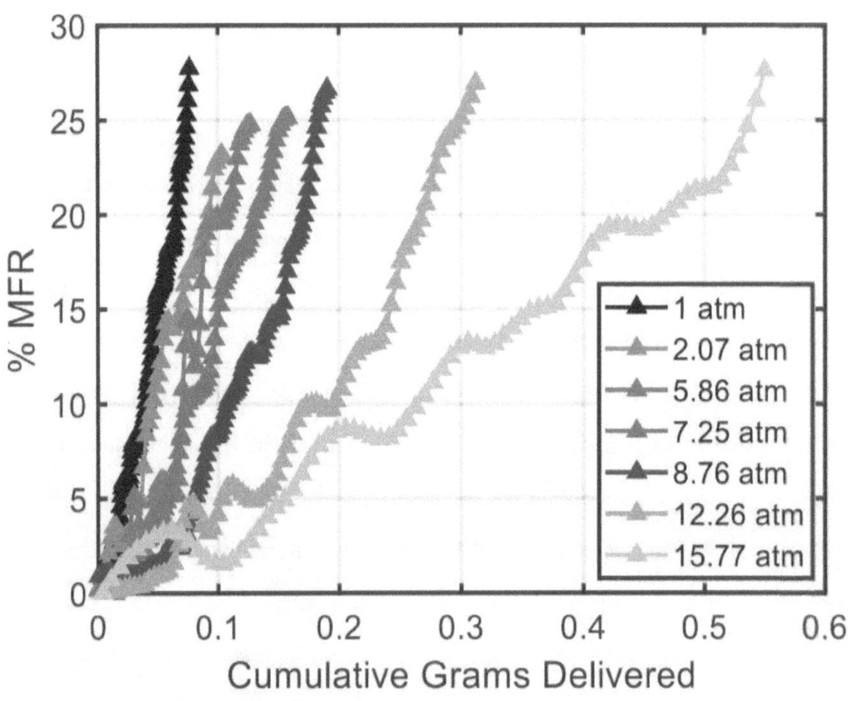

Figure 7.1: Mass flow reduction (MFR) vs cumulative grams delivered for 0-3.5 μm tests.

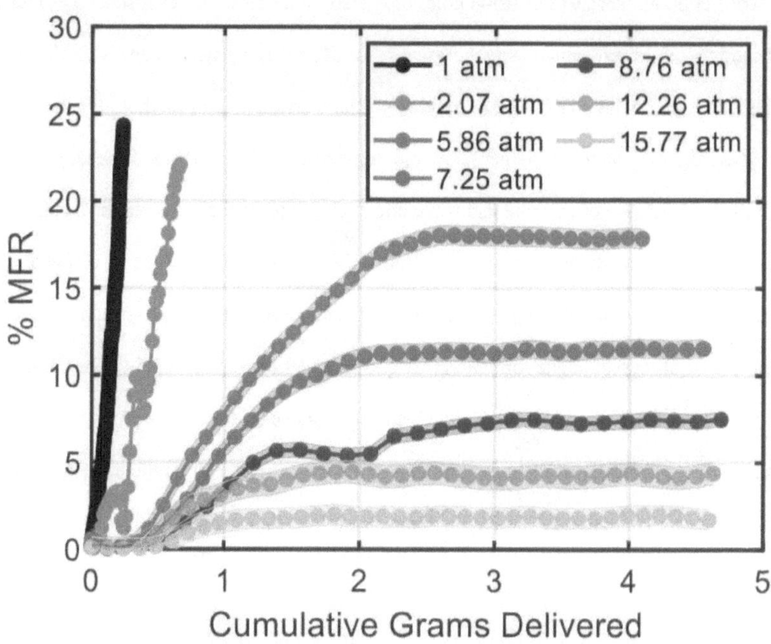

Figure 7.2: Mass flow reduction (MFR) vs cumulative grams delivered for 0-10 μm tests.

Figure 7.3 shows what the coolant side of the plate looks like after dusting with 0-10 µm at 1 atm. The area where the flow detaches from the diffuser and flows over a backward facing step before attaching to the plate is boxed in white. This region affects the flow and deposit structures in the first 6 columns, after which the deposits begin to look nominal and fully developed. At 1 atm, the deposit structures formed with both distributions build a distinct tongue shape inside the hole. Wolff et al. in their 2018 effusion hole study also found these shapes inside holes that were significantly blocked and named them the "tongue" structure. The general shape is similar for both distributions tested, with slightly more buildup on the leeward side of the hole for the 0-10 µm ARD. This might be due to the larger, slow moving particles in the distribution that ballistically impact the leeward side of the hole and settle and stick to the surface on a secondary impact. At 7.25 atm the growth on the leeward size is far more significant and builds what Wolff et al. called the "shell" structure, as it builds upward and outward and obscures the tongue structure in the hole beneath it. The obscured tongue formation is slightly smaller than in the 0-3.5 µm test at the same pressure and causes 6% less MFR. At 15.77 atm the tongue in the 0-10 µm test is completely absent, and there is only a small amount of dust lining the leeward side of the hole. The shell has also receded a bit compared to the 7.25 atm case, and the resulting deposit causes a very minor amount of blockage.

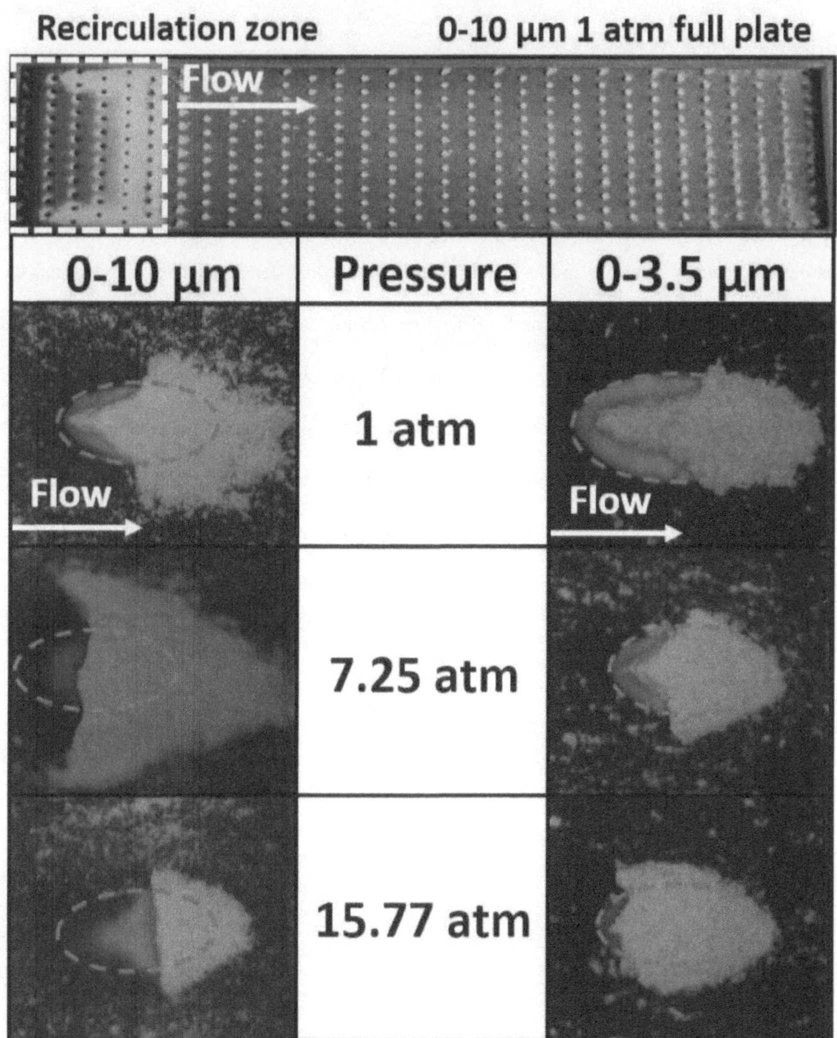

Figure 7.3: Coolant side full-plate photo of 0-10 μm 1 atm test (top). Top-down microscope images of nominal deposit structures for both size distributions at 1, 7.25, and 15.77 atm. The clean hole outline is indicated by the red dashed oval.

Chapter 8. Influence of Fluid Shear

Sacco et al. noted that as fluid pressure increases, the fluid shear that can cause removal of particles and deposit structures increases in turn. At a constant flow temperature, increasing the gas pressure causes a linearly proportional increase in the fluid density. This is a consequence of the ideal gas law, shown in Equation 8.1 The flow Reynolds number was given by Equation 3.1, and it can be shown that an increase in the density of the gas, causes a linearly proportional increase in Reynolds number. This assumes that the characteristic flow velocity remains unchanged. By fixing the pressure ratio across a cooling hole, the isentropic Mach number is fixed. A fixed Mach number and flow temperature result in a nominally constant average flow velocity. Thus, an increase in the engine pressure results in a linear increase in the flow Reynolds number.

$$\rho_f = \frac{P_f}{RT_f} \tag{8.1}$$

$$C_f = \frac{2\tau_w}{\rho_f U_f{}^2} \propto \frac{C}{Re_x{}^n} \tag{8.2}$$

Increasing the fluid Reynolds number while maintain the average flow velocity results in the thinning of the fluid and thermal boundary layers and an increase in both the wall shear stress and heat flux. For the case of canonical flows, the wall shear stress is often found through a similarity solution for the skin friction coefficient. The definition of the skin friction coefficient (C_f) as well as the typical form of the scaling between the Reynolds number (Re_d) are given in Equation 8.2. The value of "n" and "C" in Equation 7.2 are a function of the flow type as well as whether the flow is in the laminar or turbulent regime. Expanding and rearranging Equation 7.2 gives Equation 8.3, which more clearly shows the relationship between the flow pressure and the fluid wall shear. The value of "n" is less than unity in canonical flows and is equal to 0.5 for a laminar flat plate flow and 0.2 for a turbulent flat plate flow. This clearly demonstrates that an increase in fluid pressure will result in increased wall shear.

$$\tau_w \propto P_f^{1-n} T_f^{n-1} U_f^{2-n} x^{-n} \tag{8.3}$$

Particles that impact a wall are immersed in the near-wall boundary layer and are subject to drag forces that can cause them to either slide on the surface or roll and detach from the surface. Zoeteweij et al. showed that particles smaller than 100 µm tend to roll rather than slide [28]. Since particles relevant to deposition in internal cooling sections are much smaller than 100 µm, the rolling dynamic should be analyzed to determine if the rolling moment caused by fluid shear forces is large enough to overcome the counter-moment caused by adhesive forces at the contact point.

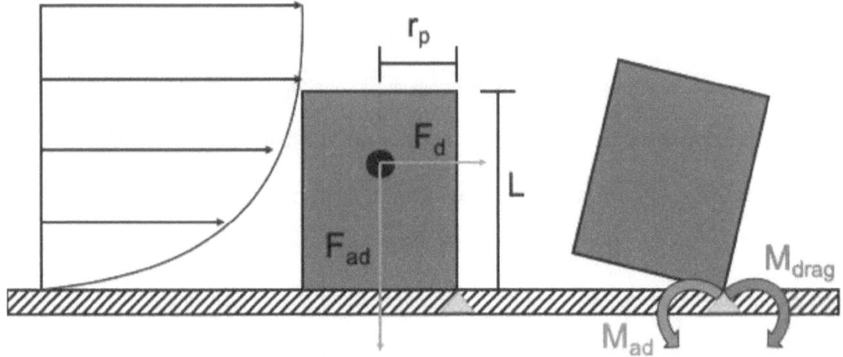

Figure 8.1: Schematic of shear and adhesion moment balance on a particle impacting a surface.

A schematic of the forces and moments due to fluid shear and adhesion on an impacting particle is shown by Figure 8.1. The schematic ignores the stored kinetic energy of the particle and the inertial moment due to the particle's inbound velocity tangential to the surface. These would combine with the fluid shear force to help the particle rebound or roll. The goal of the following analysis is to simply perform an order of magnitude analysis between the adhesion moment and the fluid shear moment over a range of pressures. Doing so for a few particle sizes relevant to small particle deposition will allow for conclusions to be made regarding the importance of fluid shear removal in the pressure trends seen in the experiments. The following analysis models the particle as a cylinder impact head on. This is the principal assumption made in the OSU Deposition Model [12]. In the OSU Deposition Model, the cylinder has equal volume to a spherical particle of the same diameter. The cross-sectional area of the cylinder is assumed equal to the frontal area of the spherical particle with the same diameter. The length of the cylindrical particle is found by dividing the particle volume by the cross-sectional area. The cylindrical particle assumption allows for simpler algebraic physics, as opposed to the numerical integration and iteration involved in spherical particle models. In reality, sand particulate is comprised of highly random and irregular shapes, so the cylindrical model is as reasonable a simplification as spherical.

The adhesion moment of each particle (M_{adh}) is calculated using Equation 8.4. The surface free energy (γ) as described in detail earlier is a constant equal to the energy per unit area required to separate two surfaces. Since most of the deposition process involves particles sticking to other particles, a γ equal to 0.4 N/m for crystalline quartz is used since this is the primary chemical in ARD and other sands. Parks reported that values of γ for quartz can range from 0.4 N/m to 11.5 N/m depending on the environmental conditions [29]. The author documents a value of γ on the low end of 0.4 N/m for quartz in the presence of saturated water-vapor, which would be a common condition in an engine due to atmospheric humidity as well as water-vapor as a combustion byproduct. Using the low-end value also provides a conservative estimate for the adhesion which is desirable for this exercise. The literature is sparse for values of γ for the other chemical constituents in ARD. The OSU Deposition Model considers an increase in the contact area during impact due to elastic and plastic deformation, but for this analysis no deformation is considered.

$$M_{adh} = \gamma A_{cont} \tag{8.4}$$

The shear drag moment is calculated using the OSU Deposition Model's implementation, which was derived from the work of Soltani and Ahmadi [30]. The model uses the cylinder in crossflow drag correlation from White [31]. The drag moment is integrated from the contact surface to the top of the cylinder and yields Equation 8.5. Equation 8.5 is multiplied by a factor of 1.7 to account for near-wall effects as done by Soltani and Ahmadi.

$$M_{drag} = \frac{\rho^3 u_\tau^4 d_p l^4}{8\mu^2} + \frac{3\rho^{5/3} u_\tau^{8/3} d_p^{1/3} l^{10/3}}{2\mu^{2/3}} \tag{8.5}$$

Equation 8.5 is evaluated for a canonical laminar axisymmetric stagnation flow as depicted by Figure 8.2, which is a representation of an impingement cooling jet. This was chosen as it is perhaps the most common cooling geometry studied in deposition, and the similarity solution allows for an algebraic estimation of the shear velocity (u_τ), which is defined in Equation 8.6. Wall shear stress (τ_w) is found from the skin friction coefficient (C_f) in Equation 8.2. The correlation for C_f is taken from the similarity solution in White and given by Equation 8.7 where F_o'' equals 1.31194. An impingement hole with diameter (d) of 0.56 mm was assumed. An average impingement jet velocity of 50 m/s and a near wall temperature of 1255 K were also assumed. All of these are common values in an engine. Since typical impingement deposits form within one to two diameters from the jet centerline, the shear moment was evaluated at x = d = 0.56 mm. The freestream velocity at this location is found using the similarity Equation 8.8.

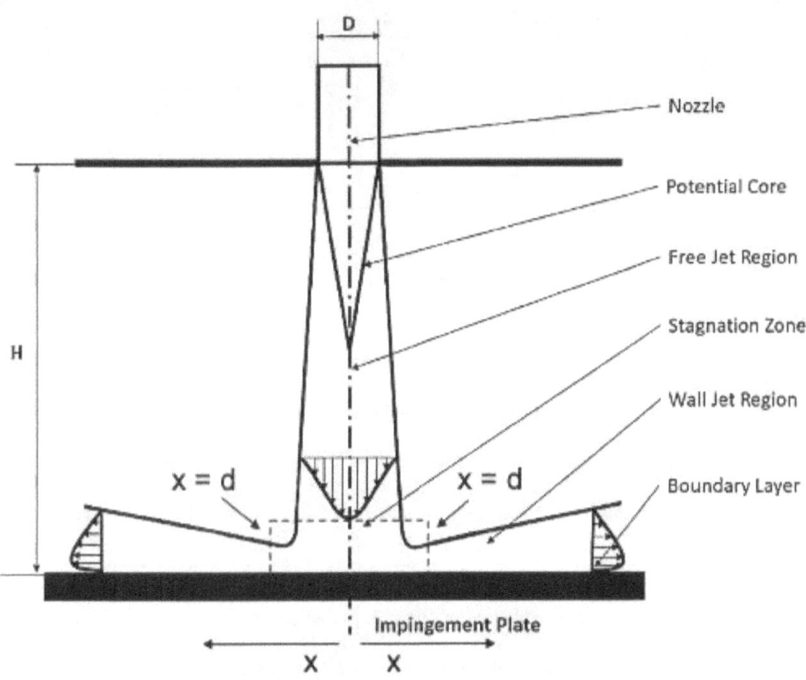

Figure 8.2: Schematic of a stagnation flow/impingement jet [32].

$$u_\tau = \sqrt{\frac{\tau_w}{\rho}} \tag{8.6}$$

$$C_f = \frac{2F_o{''}}{\sqrt{Re_x}} \tag{8.7}$$

$$U_\infty = Bx \quad where \quad B = \frac{U_{avg}}{d} \tag{8.8}$$

The ratio of the drag moment to the adhesion moment is calculated for pressure ranging from 1 to 50 atm and for 5 different particle diameters. The resulting ratios are plotted on a log axis against absolute pressure and the corresponding Re_d in Figure 8.3. The results overwhelmingly suggest that the moment due to shear drag is anywhere from 2 to 4 orders of magnitude smaller than the adhesive moment for particles relevant to deposition in internal cooling channels (< 5 µm). The drag moment is not of the same order as the adhesive moment until the particle is as large as 25 µm and at a pressure of 30 atm. This order of magnitude analysis makes several assumptions, but none that will cause smaller particles to experience shear drag moments that are orders of magnitude larger than calculated in Figure 8.3.

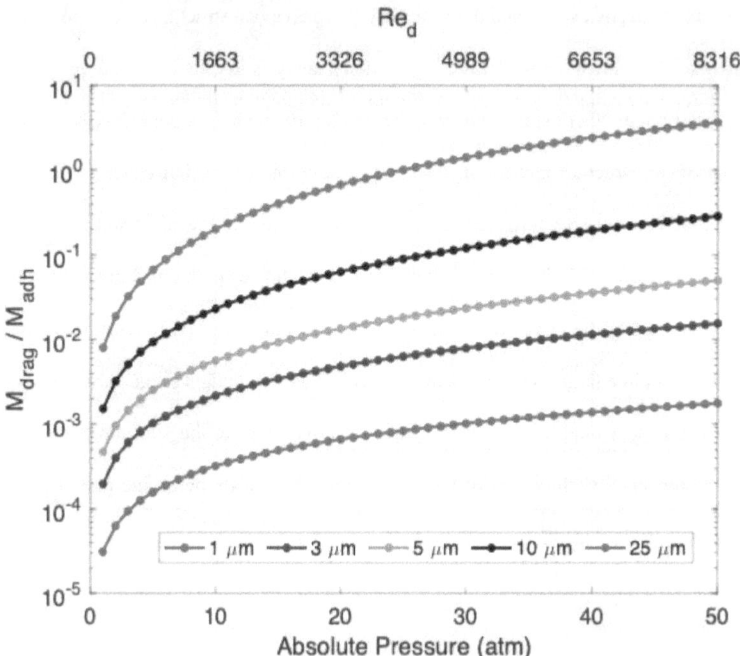

Figure 8.3: Ratio of the shear drag moment (M_{drag}) to adhesion moment (M_{adh}) vs. pressure (bottom) and corresponding Re_d (top) for five different d_p.

This assessment is necessary because an experimental evaluation of shear and adhesion forces in a turbine environment that is isolated from other significant forces is impossible. Although the conclusion is clear that shear removal of single particles is negligible for particles that have the capacity of sticking to a surface, this does not rule out shear removal of larger deposit structures on the metal surface. This is a much more difficult problem to assess analytically because of the dependence on the deposit shape and the flow-field. Large deposit structures are in non-canonical shapes that would require a high-fidelity computational analysis to calculate the drag forces and moments. Additionally, there is no guarantee that the failure mode of a deposit structure is at the contact area of the structure and the metal surface. Any one of the individual bonds between particles in contact in the structure could serve as a failure point and would result in the breakoff of part of the structure. This is a complicated fracture mechanics problem. Intuitively however, one might suppose that the large coordination number, or number of particle-to-particle contact points, might result in a larger net surface area for adhesion. This might make particles packed in a porous structure less likely to experience shear removal than in the single particle analysis. Ultimately, it appears that the effect of increasing fluid shear with increasing pressure plays an insignificant role in the deposition trends seen in this experiment.

Chapter 9. Discharge Coefficient Effects

Another effect of increasing the flow Reynolds number as pressure is elevated is the potential for an increase in the discharge coefficient of the cooling hole. The definition of discharge coefficient (C_d) is provided in Equation 9.1. It is a measure of how efficiently an orifice discharges mass flow compared to an ideal orifice. The denominator is equal to the maximum amount of mass that can be discharged if the flow is isentropic for a fixed change in pressure Δp across the orifice. An ideal hole or orifice then has a C_d of 1. In reality flows are non-ideal and have various loss mechanisms that cause the actual mass flow ($\dot{m}$) to be a fraction of the isentropic mass flow.

$$C_d = \frac{\dot{m}}{A\sqrt{2\rho\Delta p}} \qquad (9.1)$$

Rearranging Equation 9.1 to solve for $\dot{m}$ and through substitution of the ideal gas law and the definition of PR, Equation 9.2 can be achieved. This form shows that for a fixed C_d, A, T_{up}, and PR, $\dot{m}$ scales linearly with P_{up}. This is because the average flow velocity does not change, and the flow density increases linearly with pressure under the ideal gas assumption. Thus, in these high-pressure experiments, the mass flow rate will nominally scale linearly with pressure. This assumes that C_d, is constant with increasing pressure. Lichtarowicz et al. showed however that this is not necessarily the case [33]. The authors measured the discharge coefficient through a 90°

impingement hole with $l/d = 2$ and found that C_d increased with Re_d up to Re_d of approximately

10,000, at which point it became constant. The same trend has been found in effusion holes like

the ones in this experiment [34]. Increasing the fluid pressure and density in these tests increases

Re_d. There is potential then for the C_d to also increase with pressure, which would result in higher

Mach numbers and average flow velocities through the hole despite keeping PR and T fixed.

$$\dot{m} = C_d A \sqrt{\frac{2 P_{up}^2 (PR - 1)}{R T_{up}(PR)}} \tag{9.2}$$

The blockage trends of 0-10 μm ARD in Figure 7.2 could in theory be caused by increases in

C_d and average flow velocity because deposition tends to decrease at elevated gas velocities [35].

To first assess whether or not C_d is a function of pressure in the experiments, CFD on a single

effusion hole was performed for 3 different pressures (1, 5.86, and 15.77 atm). The PR across the

hole is set to 1.03 and the inlet flow total temperature is set to 950 K, both matching the experiment.

Only the absolute inlet pressure is varied. A more in-depth description of the CFD techniques and

setup employed will be discussed in Chapter 11. The steady state flow-field was solved for each

pressure, and the velocity fields were plotted on a contour for the two extreme pressures and are

shown in Figure 9.1. The average flow velocity for each was calculated ¾ of the way down the

hole and revealed that the average flow velocity increased by 22% from 1 to 15.77 atm. The corresponding C_d for the three computational pressures assessed as well as the experimental C_d values are plotted in Figure 9.2 as a function of upstream pressure (bottom) and corresponding Re_d (top). The results show that both the experimental and computational C_d values increase with pressure up to a certain pressure at which they appear to level off. The computational values are slightly underestimated compared to the experiment; however, the trends are in good agreement.

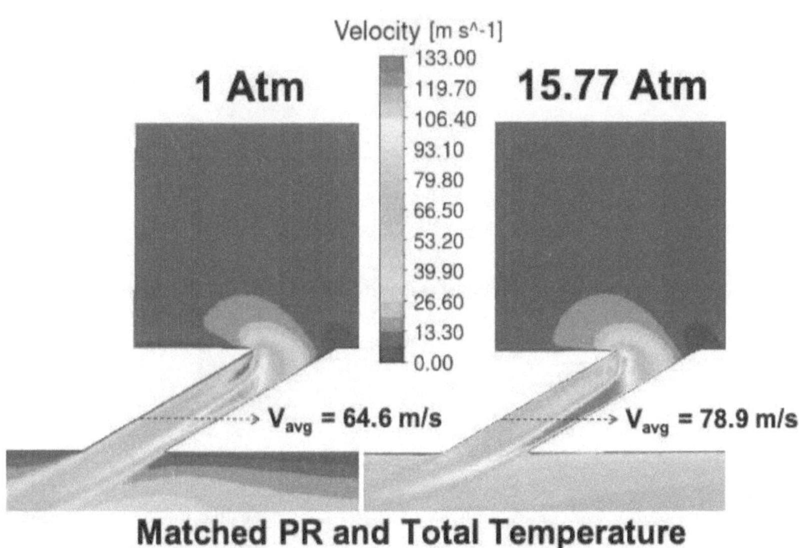

Figure 9.1: CFD velocity field contours at mid-hole for the 1 and 15.77 atm pressure tests.

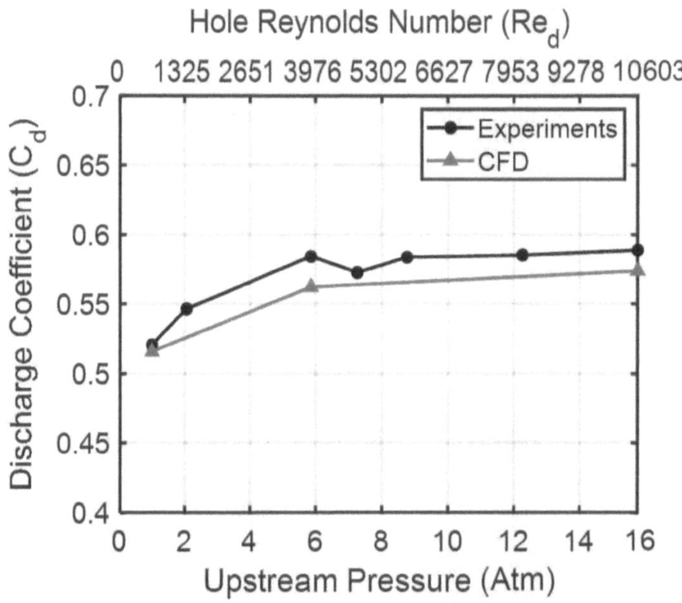

Figure 9.2: C_d vs upstream pressure (bottom) and corresponding Re_d (top) for computations and experiments.

With evidence that the C_d increases in the experiments, an experiment was run to diagnose whether the increased flow velocities at higher pressures are responsible for the reduction in the rate of deposition. The test was designed to isolate other effects that might arise with increased fluid pressure and density, such as increased drag, from the effects associated with increased flow velocities. To do this, the average velocity for the 15.77 atm test was reduced in order to match the lower velocity in the 1 atm test. To accomplish this, the upstream flow pressure was still set to

15.77 atm, but the PR was reduced to decrease the flow Mach number. Matching the flow velocity implies matching the volumetric flow rate between the two conditions. The mass flow rate then will scale linearly with pressure, so the 1 atm mass flow was simply multiplied by 15.77 to determine what mass flow rate to run for the 15.77 atm condition. This gives a mass flow of 0.035 kg/s as opposed to the 0.038 kg/s in the original 15.77 atm test. The resulting PR at 0.035 kg/s and pressure was a reduced value of 1.0255. To further validate this approach, the CFD was rerun with a mass flow inlet set to 0.035 kg/s. The resulting PR predicted by the CFD was within 1% of the experiment, lending increased confidence in the methodology.

The test was run at these conditions with 0-10 μm ARD and the MFR curve is plotted against the original 1 atm and 15.77 atm curves in Figure 9.3. The new 15.77 atm curve at reduced PR = 1.0255 (blue) shows only a slight increase in blockage with grams delivered compared to the PR = 1.03 test (red) at the same pressure. If the significant increase in blockage at 1 atm compared to 15.77 atm at PR =1.03 was due primarily to increased C_d and subsequent elevated flow velocity, reducing the PR to match the 1 atm average velocity while maintaining the upstream pressure of 15.77 atm would cause the blockage to increase and approach the levels of the 1 atm test (black). There appears to be some minor effect of C_d since the blue curve does move slightly toward the black curve, however, it is evident that some other physics is the primary mechanism causing dramatic decreases in blockage at elevated pressures. Although changes in C_d may be more influential in deposition testing in alternate geometries, the effusion trends shown in Figures 7.1 and 7.2 must be caused by an alternate mechanism.

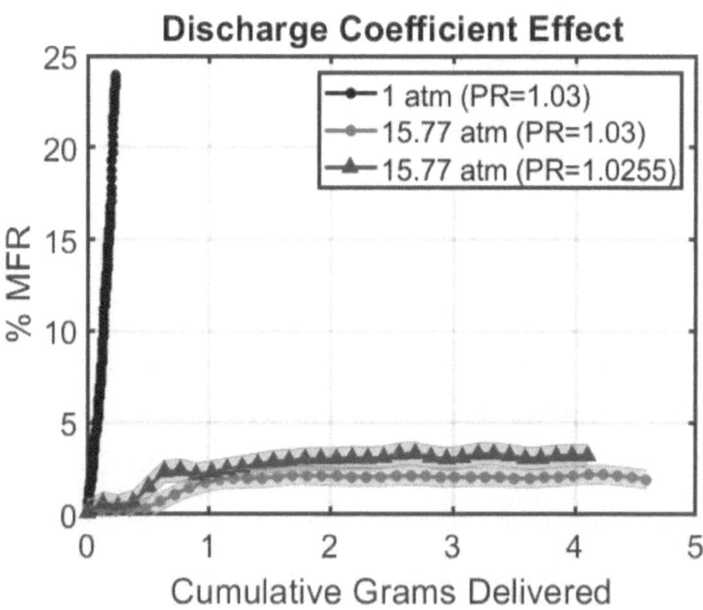

Figure 9.3: Test MFR vs cumulative grams delivered results analyzing the role of changing C_d with pressure.

Chapter 10. Role of Particle Drag on Sticking

With two of the mechanisms shown to be unlikely to reduce deposition and ensuing blockage in the effusion holes in this experiment, the third mechanism emerges as the most likely candidate. Sacco et al. showed that at elevated absolute pressure particles experienced increased fluid drag that can enable dust to maneuver through cooling holes and passages more readily. The prior mechanisms investigated revolved around an increase in flow Reynolds number. The particle Reynolds number was introduced in Chapter 3 but is recast for clarity in Equation 10.1. The increase in flow density causes a linear increase in particle Reynolds number (Re_p) as it does in Re_d. The following analysis will start with making a spherical particle assumption, which is a very common simplification made in particulate studies.

$$Re_p = \frac{\rho_f(U_f - U_p)d_p}{\mu_f} \tag{10.1}$$

$$F_d = 3\pi\mu_f d_p(U_f - U_p) \tag{10.2}$$

$$St_k = \frac{\rho_p d_p{}^2 U_f}{18\mu_f L} \tag{10.3}$$

$$St_{keff} = \varphi(Re_p)St_k \tag{10.4}$$

When a spherical particle has a Re_p less than 3, Stokes Law applies. This analytical law states that the particle drag (F_d) is 2/3 viscous drag and 1/3 pressure drag. The net drag on particles in this regime is referred to as Stokesian drag and is shown in Equation 10.2. The drag is notably not a function of the fluid density, so particles in Stokes flow do not experience an increase in drag as flow pressure and density increase. The Stokes number was also described in Chapter 3 but is also provided again for convenience in Equation 10.3. This parameter compares the inertia of the particle to the Stokesian drag of Equation 10.2 and is used to characterize how likely particles are to follow flow streamlines. A particle with St_k less than unity is likely to aptly follow the streamlines, while a St_k larger than unity will tend to act more ballistically.

In high-pressure flows, the elevated fluid density results in Re_p larger than 3 and the particle drag behavior departs from the Stokesian drag law of Equation 10.2. Figure 10.1 plots the drag coefficient (C_D) of a sphere versus Re_p under the Stokes Law assumption. It also plots the C_D curve collected from experimental measurements over the wide range of Re_p. The deviation of the curves occurs around $Re_p = 3$ as expected, and at large Re_p the "actual" drag is higher than what would be predicted using Stokes Law. At $Re_p > 100$, the difference is more than an order of magnitude. In the non-Stokesian regimes, the fluid density re-emerges as a variable that affects the particle drag, and thus increases in pressure that occur at $Re_p > 3$ will result in additional drag. The traditional Stokes number of Equation 10.3 is not strictly valid in these circumstances, as it underestimates the drag. The effective Stokes number defined in Equation 10.4 was derived to more accurately

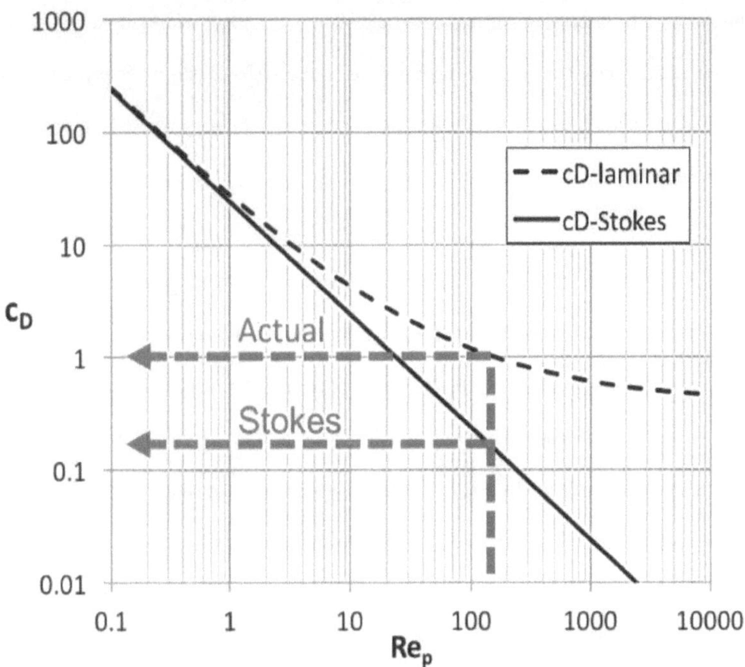

Figure 10.1: Drag coefficient (C_D) versus Re_p for smooth spheres under Stokes Law and actual experimental drag measurements.

characterize the flow following tendencies of particles at high Re_p. This parameter has been used by many authors including Israel and Rosner [36] who used it to characterize capture efficiency curves in aerosol collectors. Particles suspended in high-pressure flow-fields will have an effective Stokes number that is lower than the traditional Stokes number. This will lead to particles that follow flow streamlines more readily than they do in low pressure environments. This is likely to reduce the number of impacting particles. Particles that still impact will likely do so at reduced angles and have a larger velocity component parallel to the surface. This gives particles a greater ability to roll and detach. The net effect is that the net capture efficiency would be reduced as seen in these experiments.

While it's clear that even under the assumption of spherical particles increasing pressure can result in additional particle drag, consideration must be given to the validity of the spherical assumption over a range of pressures. Spherical particle models are convenient, but common dust particles are highly non-spherical. Figure 10.2 shows the irregular shapes in ARD as viewed with a scanning electron microscope. There are numerous ways that the deviation from a sphere is quantified depending on the application, but the most common is the shape factor. The 3-D shape factor (ϕ), or volume shape factor is defined in Equation 10.5. In this equation, "S" is the surface area of the non-spherical particle, while "s" is the surface area of a spherical particle with volume equal to the non-spherical particle.

Figure 10.2: High-resolution image of ARD particles taken with a scanning electron microscope.

$$\phi = \frac{s}{S} \tag{10.5}$$

Non-spherical particle drag has been an active area in aerosol research for decades, and most models make use of the shape factor to predict how the drag of an irregularly shaped particle deviates from the spherical drag curve. Perhaps the most widely utilized drag correlations in commercial CFD particle tracking applications are those of Haider and Levenspiel [37]. The C_D correlation is a function of ϕ and Re_p and is provided by Equation 10.6.

63

Figure 10.3 shows the C_D curves using this correlation for 6 values of ϕ, including $\phi = 1$ for a perfect sphere, over a range of Re_p. The Stokes drag is also included in the plot. A typical range of Re_p in deposition studies for 0-10 μm particles over a wide range of absolute pressures is depicted inside the transparent yellow box. At lower Re_p typical of low-pressure applications, it is seen that the various C_D curves appear to collapse. This appears to begin at roughly $Re_p = 10$. This has long provided justification for making the spherical assumption in particle modeling, as it will predict approximately the same drag as the more complicated non-spherical model and has the benefit that ϕ does not need to be determined.

$$C_D = \frac{24}{Re_{sph}}\left(1 + b_1 Re_{sph}{}^{b2}\right) + \frac{b_3 Re_{sph}}{b_4 + Re_{sph}} \tag{10.6}$$

where

$$b_1 = \exp(2.3288 - 6.4581\phi + 2.4486\phi^2)$$

$$b_2 = 0.0964 + 0.5565\phi$$

$$b_3 = \exp(4.905 - 13.8944\phi + 18.4222\phi^2 - 10.2599\phi^3)$$

$$b_4 = \exp(1.4681 + 12.2584\phi - 20.7322\phi^2 + 15.8855\phi^3)$$

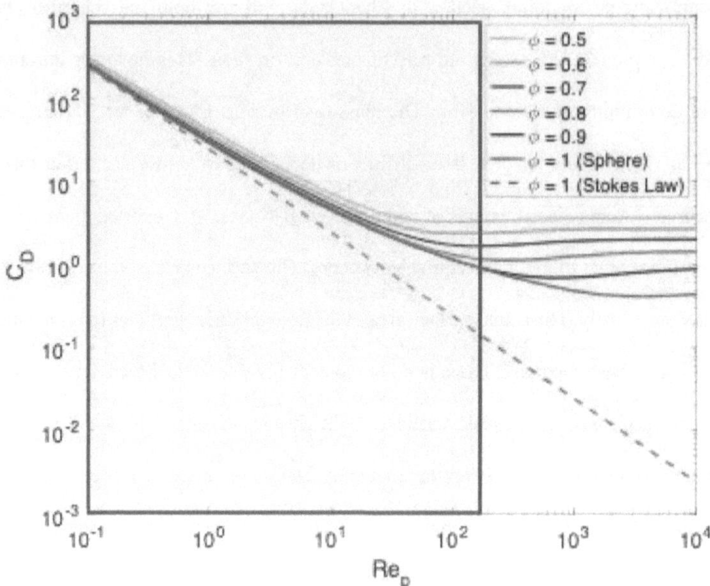

Figure 10.3: C_D curves using Haider and Levenspiel correlation for various particle shape factors (ϕ) compared to spherical C_D curves.

At the higher Re_p, however, there is noticeable separation between the trends. Considering the y-axis is a logarithmic scale, this means that depending on the value of ϕ, drag on non-spherical particles at high-pressure can be significantly higher than what is predicted using spherical drag correlations. This is a relevant realization because if the effect of pressure on deposition is indeed caused by elevated drag effects, those differences may be totally obscured or at least underestimated if modeled under the popular spherical assumption.

The computational modeling section in Chapter 12 will use both the spherical and non-spherical drag approaches to assess the role of pressure on drag. This however introduces the difficulty of determining the value of ϕ. The mass distributions provided by PTI are measured using either a Coulter counter or a laser-diffraction technique with Mie scattering theory and provide accurate volume measurements of non-spherical particles. The reported diameter for each measured particle is that of a volume equivalent sphere. The surface area of the equivalent sphere (s) can be deduced easily. The actual surface area of the non-spherical particles (S) is not measured or provided using these methods. There is no obvious or reliable optical technique to accurately measure the surface area of microscale particles. In the absence of reliable methods, the 3-D shape factor is often assumed to be closely approximated by 2-D shape factor measurements. These can be easily obtained from 2-D microscopy images. The surface shape factor, often referred to as circularity, is the 2-D equivalent of the 3-D volume shape factor. The circularity (ϕ_{circ}), defined in Equation 10.7, is defined as the squared perimeter of the circle with area equal to the projected area of the particle (A), divided by the true squared perimeter (P^2) of the non-spherical particle.

$$\phi_{circ} = \frac{4\pi A}{P^2} \tag{10.7}$$

To provide an estimate for the shape factor, the circularity for 0-10 µm ARD was measured using a series of microscope images of sparse dust particles. A nominal microscope image can be seen in the top image in Figure 10.4. The series of images was analyzed using MIPAR image analysis software. The software identifies individual particles within a given image and outputs a binary (black and white) version of the image. The corresponding binary image for the top image in Figure 10.4 is shown directly below. A pixel to length calibration for the image is provided and the software calculates shape measurements of the binary image such as circularity, elongation factor, aspect ratio, and area equivalent diameter. The series of images are processed as in a batch and the particle statistics are output to a readable file. 28 images were taken, and 17,130 individual particles were identified.

Figure 10.5 shows a scatter plot of the resulting circularity values for the imaged particles. A black line is drawn through the running mean circularity. The plot is seen to be quite sparse for effective diameters larger than 5 μm, but the mean circularity appears to decrease with increasing diameter, a trend also found by other authors [38]. The circularity appears to asymptote somewhere around 0.6, but particles more influential in the sticking and buildup (< 5 μm), the value is closer to $0.65 - 0.7$. Ghanem et al. [39] reported a typical shape factor for sand of 0.5. Sand shares quartz as its primary constituent, but the exact chemical composition and the region and method collection and preparation is likely to affect the measured value. For the purposes of the CFD modeling in this study, a value of 0.65 is assumed.

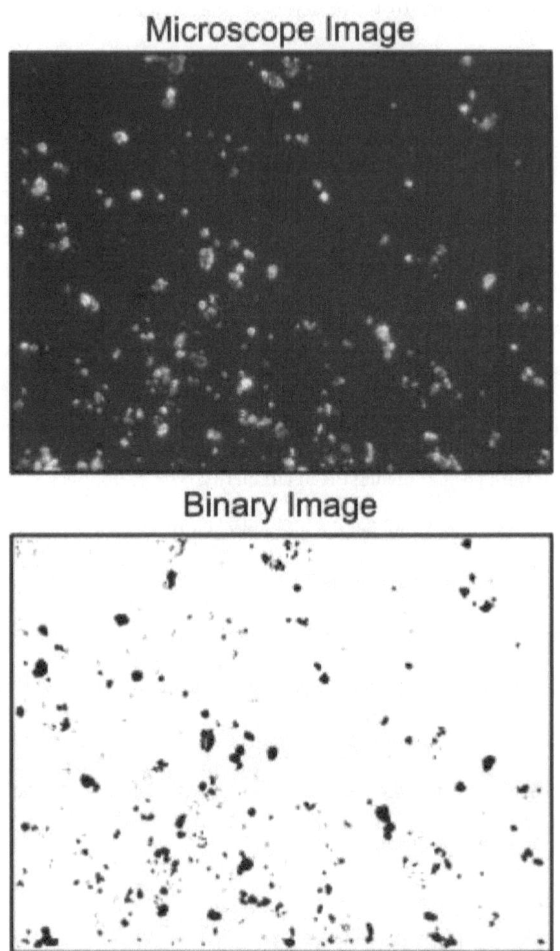

Figure 10.4: Microscope image of 0-10 μm ARD particles (top) and binary image processed using MIPAR software (bottom).

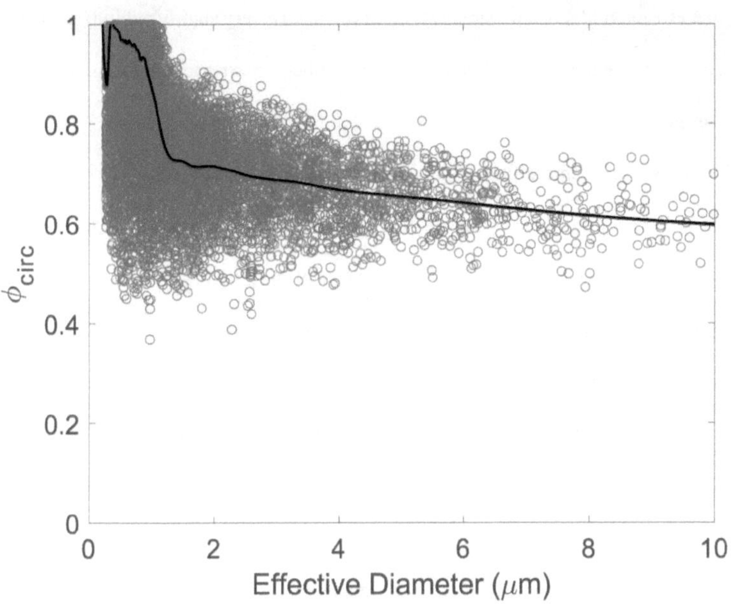

Figure 10.5: Circularity (ϕ_{circ}) vs effective particle diameter for 0-10 ARD particles.

Chapter 11. Role of Particle Drag on Erosion

The role of added drag in reducing the likelihood of both impact and sticking is logical, but this is unique to small particles within a dust distribution. The work of Wolff in 2018 et al. [19] suggested that erosion of deposit by larger particles can have a very significant impact on effusion hole blockage, so the degree to which erosive physics affect the experimental results at pressure deserves examination.

Wolff et al. used the same effusion plate that is used in this study and did a comprehensive analysis of how different size distributions of dust affect the blockage. Among the primary findings was a confirmation of the earlier findings of Whitaker et al. [20] who suggested that particles smaller than approximately 3.5 μm are responsible for the sticking and development of deposits, while particles larger than approximately 5 μm have the capacity to erode and remove existing structures. This erosive ability was attributed to the higher kinetic energy of more massive particles, which scales with the cube of the particle diameter. Figure 11.1 shows microscope images of two deposit structures from the test of Wolff et al. The authors first delivered 0-10 μm ARD to the effusion plate until the MFR level for the plate reached the same 25% threshold used in the experiments in this study. A nominal resulting deposit structure is shown in the left image (a) in Figure 11.1. The authors subsequently delivered the same amount of total mass of a much larger distribution of 10-20 μm ARD. The result after that delivery is shown in the right image (b).

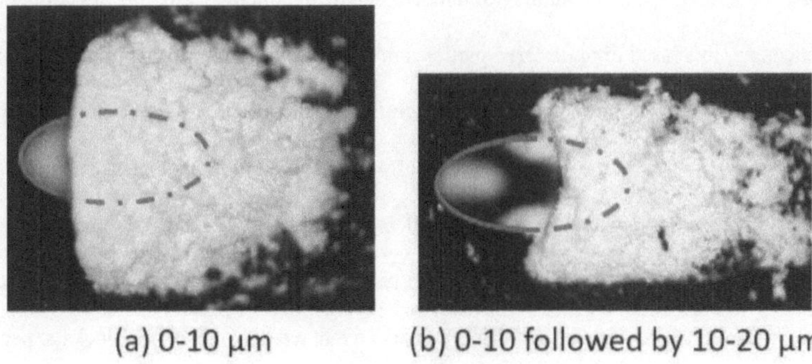

Figure 11.1: 150° effusion hole deposit structures from Wolff et al. showing erosive capability of larger particles. Flow enters the holes from left to right in the images.

The shell structure that builds over and obscures the hole clearly receded, and the "tongue" structure inside the hole was cleared as well. This tongue formation is thought to be responsible for the majority of hole blockage as it juts into the cross-section of the hole and reduces the minimum flow area significantly. The erosion of the tongue by the larger distribution of ARD caused approximately a 10% recovery in mass flow in this experiment. The experiments by Wolff et al. were all performed in a low-pressure facility, with no variation in absolute pressure across tests. In the current experiments in the HPDF increasing pressure and drag on the particles larger than 5 µm may alter the trajectories in a way that enhances the amount of erosion that occurs at elevated pressure. The 0-10 µm distribution obviously contains particles smaller than 3.5 µm and larger than 5 µm. The deposition results for this distribution are challenging to assess because the influence of a reduced effective Stokes number causes competing effects. Small particle trajectories are affected, and the number of impacts and the likelihood of sticking are both reduced. Simultaneously, increased drag on large particles may be allowing for more high energy impacts and increased erosion. To assess the relative magnitude of the role that particle drag has in these two ranges of particle sizes, the 0-3.5 μm ARD tests were added. This smaller distribution of dust should eliminate erosion physics, isolating the effect of drag on small particle deposition.

Comparing the blockage trends between the two distributions allows for insights into how erosion is affected by pressure in the 0-10 μm dust. To make this comparison, blockage per gram (BPG) is reintroduced as a useful metric. BPG is defined by Equation 11.1, and is simply the MFR divided by the total grams of dust (g_f) required to achieve that MFR. For tests that were terminated because the 25% MFR level was reached, g_f is determined simply by the total dust delivered. For tests where MFR never reached 25% and all of the dust was delivered, g_f is defined at the point where the MFR levels off. The rest of the dust delivered after that point is not contributing to blockage and thus would reduce the BPG in a non-physical sense.

$$BPG = \frac{\%MFR}{g_f} \tag{11.1}$$

$$BPG_{<3.5} = \frac{\%MFR}{g_f < 3.5\ \mu m} \tag{11.2}$$

This metric can be thought of as the average rate of blockage during the experiment. Referring back to the MFR results of Figures 7.1 and 7.2, it is evident that the blockage is not linear throughout the duration of the dust delivery. BPG is nonetheless useful in comparing how aggressively deposition occurs on average for various test conditions. BPG also suggests that the rate at which particles are delivered to the test article is negligible and only the total amount of delivered dust is relevant to predicting the level of blockage that will occur. Whitaker et al. [20] varied the dust loading rates in a double wall cooling experiment and found that the loading could be varied by several orders of magnitude without affecting the final amount of deposition, as long as the total dust delivered by the end of the experiment was held constant.

Figure 11.2 plots the BPG for both the distributions of ARD over the range of pressure tested. The plot is consistent with the MFR results shown in Figures 7.1 and 7.2, but the use of BPG in this manner could lead to confusion or incorrect conclusions about the physics causing these changes for the two distributions respectively. Wolff. et al also introduced a modified version of BPG that considers only the amount of dust delivered smaller than 3.5 μm ($BPG_{<3.5}$). $BPG_{<3.5}$ is defined in Equation 11.2 and divides the MFR by the mass of dust delivered that is below the sticking threshold of 3.5 μm. 96% of the total mass is below this threshold for the 3.5 μm distribution, whereas for the 0-10 μm ARD only 46% is below 3.5 μm. This is a logical modification since the blockage in the numerator is thought to be caused by small particles that stick and build deposits inside the holes. If the particles larger than 3.5 μm played no role in the test but accounted for a significant portion of the delivered mass, including that non-participating mass would drive down the BPG in a way that is non-physical. By normalizing in this manner, the physics that affect the rate in which small particles cause blockage can be assessed correctly. Wolff et al. showed that this normalization caused the BPG for various distributions of dust to collapse within $\pm$ 30%.

Figure 11.3 replots the data using this metric. The 0-3.5 μm trend decreases non-linearly with pressure initially but eventually appears to linearize. 0-10 μm has an even steeper initial drop off in $BPG_{<3.5}$ and eventually asymptotes to $BPG_{<3.5} \cong 0$. The $BPG_{<3.5}$ at 1 atm is approximately 40% lower for the 0-10 μm than for the 0-3.5 μm. The experimental flow conditions are exactly matched, indicating that the reduction in $BPG_{<3.5}$ is caused by large particles in the 0-10 μm dust eroding deposits while smaller particle build them. Looking at the decrease in $BPG_{<3.5}$ from 1 atm to 2.07 atm for 0-3.5 μm the reduction is approximately a factor of 1.5. For 0-10 μm, the $BPG_{<3.5}$ is reduced by roughly a factor 3. The decrease in blockage with this initial increase in pressure is prominent for both distributions but has nearly double the impact for the 0-10 μm test. A reasonable theory to explain the results is that the increased drag reduces the amount of sticking by particles in the 0-3.5 μm range. This effect is present in both distributions that are tested. For the 0-10 μm, there is an additional reduction in blockage with pressure due to the enhanced erosion of deposits by large particles, causing a steep non-linearity.

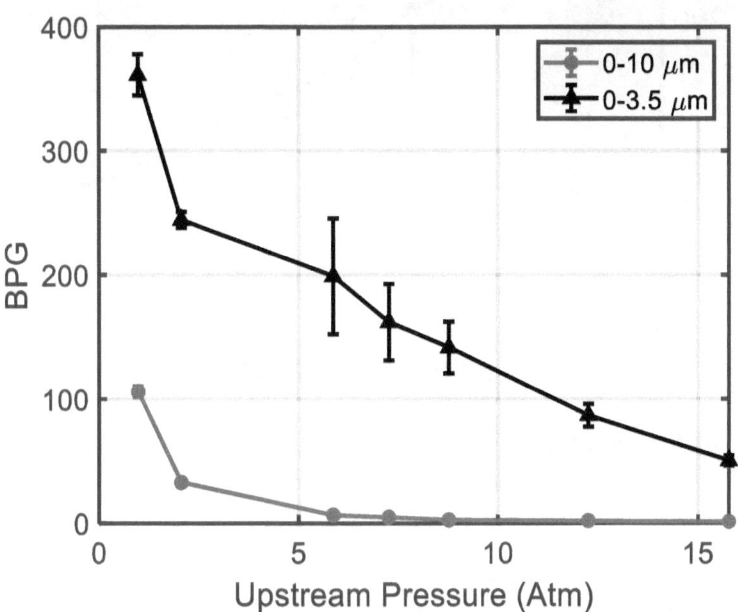

Figure 11.2: Blockage per gram (BPG) of dust vs upstream pressure for both distributions.

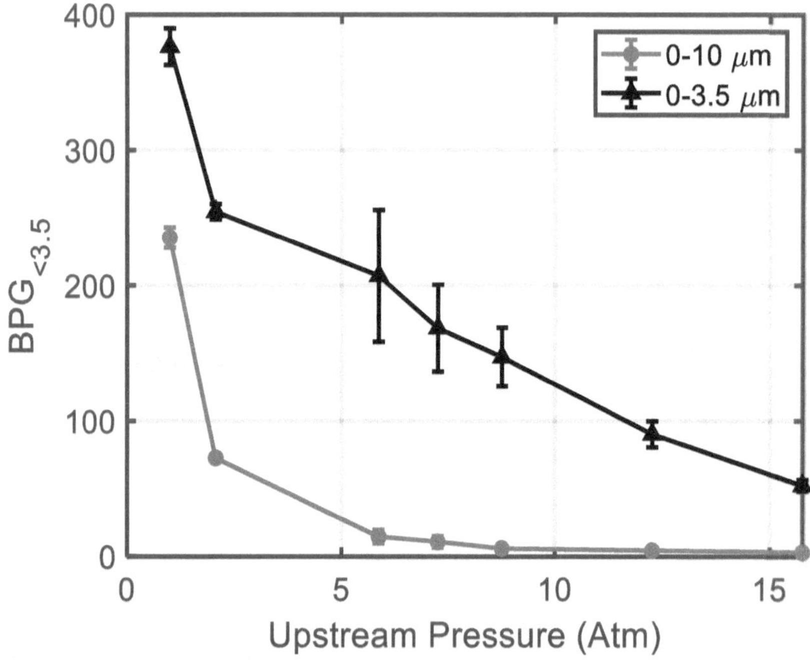

Figure 11.3: Blockage per gram of dust below 3.5 μm (BPG$_{<3.5}$) vs upstream pressure for both distributions.

To verify that erosion is an important mechanism in these experiments, a test similar in nature to that of Wolff et al. was performed. 0.6 g of 0-3.5 μm was loaded onto the front of the conveyor belt, followed by 2.2 grams of 0-10 μm. The facility was pressurized and heated to the conditions of the highest-pressure case (15.77 atm). This pressure was chosen because the 0-10 μm dust showed no significant blockage at this pressure, ensuring that the small particles in the distribution would cause minimal blockage and the erosion by large particles would be isolated.

Figure 11.4 shows the MFR vs cumulative grams delivered during the test, with the two segments clearly distinguished. The plot refers to this sequential delivery as the "combo" test, and the MFR is plotted in black. The first portion of the test consisting of the delivery of 0-3.5 μm is denoted by triangles, while the 0-10 μm section is plotted with circular markers. The original test data at 15.77 atm from Figure 7.1 and 7.2 are plotted with red triangular and circular markers respectively and labeled the "solo" tests. The goal of the 0.6 gram of 0-3.5 μm delivery was to achieve approximately 25% MFR, however the blockage overshot, and the value reached 36%. The subsequent delivery of 0-10 μm caused erosion of the deposits built in the first part of the test and caused a 25% mass flow recovery. The erosion did not cause enough of a recovery to completely clear out the blockages and result in the same final blockage as the 0-10 μm solo test (red circles). This emphasizes that the simultaneous sticking and erosion in the solo tests is critical to the process. The bottom images in Figure 11.4 show nominal deposit structures for the solo 0-3.5 and 0-10 μm tests (left and middle) as well as the combo test (right). The tongue structure that caused blockage in the first segment of the combo test was completely eroded during the second segment, causing the recovery in mass flow and a resulting structure that looks nearly identical to the solo 0-10 μm final deposit shape, which has a shell on the leeward side but no tongue.

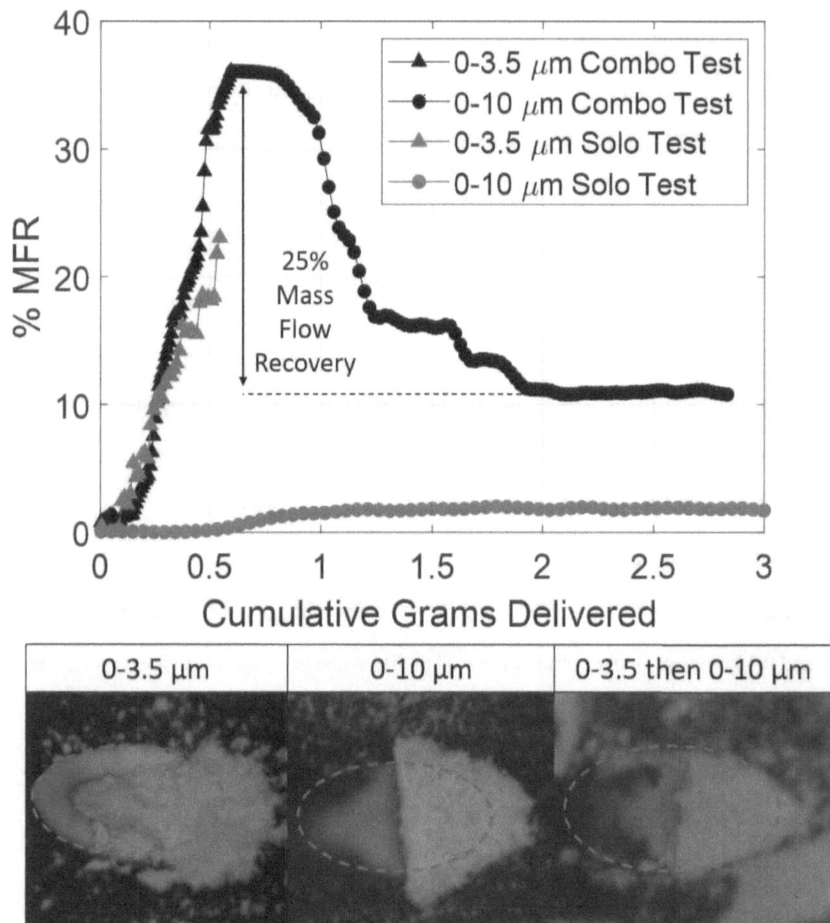

Figure 11.4: MFR vs cumulative grams at 15.77 atm for sequential 0-3.5 µm and 0-10 µm delivery (top). Microscope images of nominal deposit structures of the solo tests (left, middle) and combo test (right). Each clean hole is outlined in red.

Chapter 12. Computational Investigation Setup

The proposed theory regarding the role of increasing particle drag with pressure and the competing effects it might induce between the smaller and larger particles in the 0-10 μm ARD tests is vetted computationally. The objective is to solve the steady flow-fields at the extreme pressure of the experiment, track particles through both flow-fields, and run a mesh-morphing procedure to predict the growth of the deposit at both pressures using the OSU Deposition Model for making bounce or stick determinations. The CFD and particle tracking will in theory predict changes in particle trajectories due to changes in drag with the increased pressure. The OSU Deposition Model should be capable of predicting changes in sticking and capture efficiencies to a reasonable degree of accuracy. With a reliable routine to move the mesh as particles stick to the walls of the effusion plate, the reduction in mass flow can be calculated as a function of computational grams delivered and compared to the experiments.

To model the entirety of the effusion plate would be impractical as the grid and computational expense would be prohibitive. For the goals of this study, a single effusion hole is sufficient to draw conclusions. The hole in the center of the plate was chosen. The entire width of the diffuser is modeled as pictured in Figure 12.1. Three pairs of translational periodic interfaces are used to model the interaction between the flow of the adjacent holes both laterally and vertically to the modeled hole. These interfaces are colored yellow in Figure 12.1. Two pressure outlets are used, one for the crossflow exit and one for the main flow exit (colored orange). A stagnation pressure

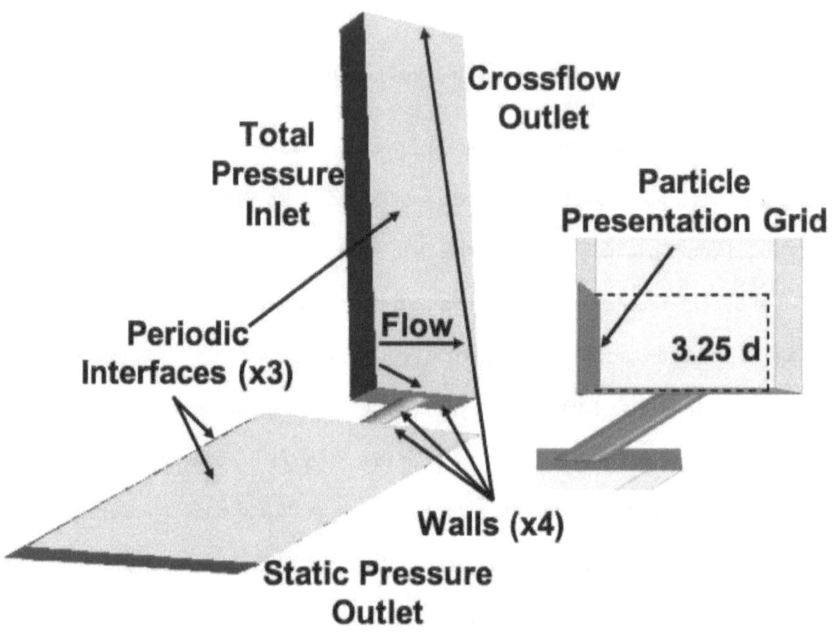

Figure 12.1: Computational domain (left) with boundary condition types labelled. Particle injection grid on the inlet shown in green (right).

inlet (colored purple) is used with a defined absolute total pressure and temperature matched to the experimental and with a defined inlet flow angle.

An unstructured tetrahedral mesh was created in ANSYS Workbench. Three grid refinement levels were created and solved at both pressures simulated to ensure grid independence. The mesh is shown on a plane aligned with the major axis of the hole inlet and cutting through the domain

80

in Figure 12.2. Added refinement can be seen in the hole area where the flow gradients are highest. A close-up view of the expansion corner at the hole inlet is also provided in the figure to give a closer look at the prism layers in the wall region. The final mesh has a total cell count of 595,357 cells and 5 prism layers with an expansion rate of 1.3. In the 1 atm case, this results in an average non-dimensional wall distance (y^+) in the hole of 1.05. In the high-pressure case, the average y+ is approximately 15 and wall functions are used in part to resolve the near wall turbulence. While it would be ideal to resolve the near wall at high-pressure, this would require very thin near wall prism layers that become unstable in the dynamic meshing procedure.

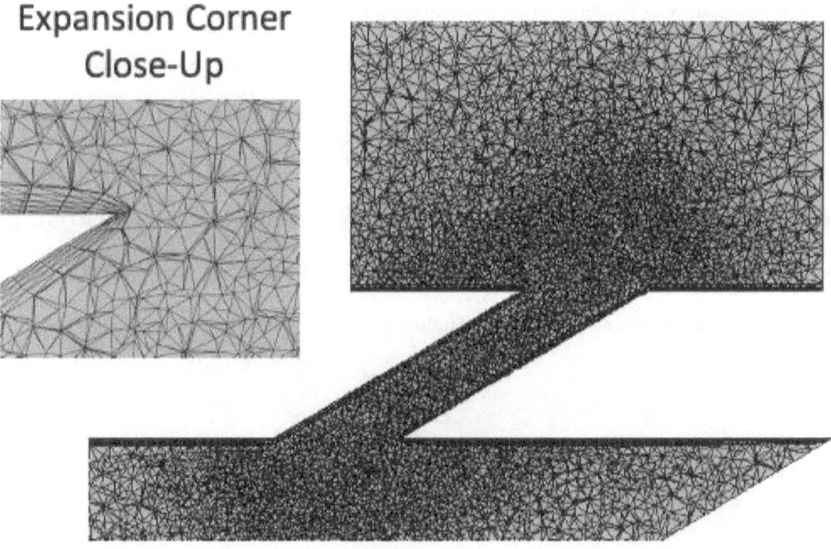

Figure 12.2: Computational mesh viewed on a plane through the hole centerline. (right). Close-up view of expansion corner with near wall prism layers (left).

A segregated pressure-velocity solver was used in addition to a segregated energy equation. A SIMPLEC algorithm was employed and all equations were discretized with a second order accurate formulation. A steady state k-ω SST RANS turbulence model was chosen as it typically offers the best mix of accuracy in wall-bounded shear layers like the ones that develop in the effusion holes, as well as in the free-shear layers that develop where the jet exits the hole. The inlet total pressure was set to either 1 or 15.77 atm depending on the case tested. To match the experiments, an inlet total temperature of 950 K is also prescribed. The turbulence intensity was determined from a channel flow correlation [40]. At 1 atm it was set to 5.4%, while at 15.77 atm it drops to 3.7%. The inlet flow angle is set to a 6° angle from normal to the inlet (inclined toward the hole). This angle was determined from a prior computation where one full row of 16 effusion holes on the plate was modeled to assess how the flow approaches the holes as a function of distance from the backward facing step of the diffuser. The primary outlet is a static pressure outlet with a pressure set to establish a PR of 1.03 across the hole. The crossflow outlet static pressure was set to match the pressure drop from the inlet to the crossflow outlet found from the 16-hole simulation. The plate and hole walls were approximated as isothermal boundaries and set to match the thermocouple measurement on the outer surface of the plate as reported in Table 5.1. The velocity contours for the steady state flow solutions at each pressure were presented in Figure 9.1.

Chapter 13. Particle Tracking and Mesh Morphing Procedure

After solving the flow-fields, particles are injected and tracked through the domain to begin the first iteration of the mesh morphing procedure. The right image in Figure 12.1 shows the injector array in green just inside the inlet of the domain. The particle array spans the width of the inlet, but only extends vertically 3.25-hole diameters from the plate surface. The particles furthest from the plate will have enough energy to pass from the inlet to the crossflow exit without ever coming close to interacting with the plate. There is no need to model particles any further from the plate as they simply do not have an opportunity to enter the effusion hole that is modeled. The array contains 10,000 evenly spaced injector points. In an experiment, millions of particles are delivered, but it is impractical and unnecessary to attempt to inject that many particles. Instead, a statistical approach is implemented. 10 discrete particle diameters are selected across the 0-10 μm range (1, 1.5, 2, 2.5, 3, 3.5, 4, 5, 7, and 10 μm). Changes in the aerodynamic and sticking characteristics with size tend to be more significant at smaller diameters, so 8 of the 10 diameters selected are ≤ 5 μm. For each size studied, 10,000 particles of that diameter are injected into the flow and tracked. The number of particles injected is increased until the capture efficiency (η_c) calculated in each wall cell is deemed statistically converged. It was determined that for this study η_c was insensitive to further increases in the number of particles delivered at 10,000. Particles are assigned an initial velocity vector and temperature that equal the flow conditions at that location.

The particles are injected and tracked using ANSYS Fluent Discrete Phase Model (DPM). Each particle is tracked in a Lagrangian frame of reference, using the general forms of Equations

2.1 and 2.2 to calculate the particle kinematics and heat transfer. The physics assume that the flow is one-way coupled, meaning that the flow affects the way that the particles move throughout the domain, but the dust is dilute enough that it does not have an impact on the fluid solution. As discussed in the experimental procedure section, the delivery rate of the particles was controlled to ensure this assumption was valid. The intensive particle properties required to correctly predict the particle trajectories and temperatures are the particle density (ρ_p) and specific heat (c_{pp}). The values for ARD used in the simulations are provided in Table 13.1 along with the mechanical properties needed for the sticking predictions.

The forces acting on the particle are modeled, and in this study the drag force, Saffman lift force, and turbulent dispersion were considered. As described previously, the effect of making a spherical drag assumption is of interest at high pressure. For each pressure, two mesh morphing routines are considered: one using a spherical drag correlation by Morsi and Alexander [41], and one using the non-spherical drag correlation of Haider and Levenspiel [37] with $\phi = 0.65$. The Saffman lift is calculated using the default Fluent formula of Li and Ahmadi [42] and captures the lift force generated on micron sized particles immersed in a shear layer. The turbulent dispersion approach native to Fluent is called a Discrete Random Walk model. Particles in the experiment experience random forcing due to fluid turbulence. When modeling the flow with a steady RANS solver, a time-averaged flow is calculated and thus the drag effects on particles due to the mean behavior of the bulk fluid are captured while the turbulent effects are ignored. The discrete random walk provides the ability to simulate turbulence in a steady flow and "nudge" particles off their mean path by using the local RMS fluctuating components of the flow-field and converting those into an equivalent body force on the particle. In the case of isotropic turbulence, as is assumed in the k- ω class of turbulence closure models, the velocity fluctuations are equal in every direction at a given point and can be found from the local turbulent kinetic energy (k) as shown in Equation 13.1. Each of the fluctuations is multiplied by a Gaussian random number to characterize the random fluctuation of that eddy. The DRW model gives reasonable results in inertial dominated flows but has the tendency to move particles into regions of low-turbulence in diffusion-dominated flows. Forsyth et al. [43] showed that in typical deposition applications including pipe flows, this can cause a very unrealistic overprediction in the capture efficiency as particles are become concentrated in the viscous sublayer.

The deposition community is actively developing Continuous Random Walk (CRW) models that simulate a more realistic interaction between the turbulent eddies and particles in diffusion dominated flows through solution of the Langevin equation [43][44]. While the CRW models undoubtedly provide needed accuracy for common engine flows, they add more computation time because the integration time required to resolve the turbulence approaches the smallest Kolmogorov time scales. That makes the DRW a more suitable model for the purposes of this study. Additionally, the tongue structures that form in the effusion holes are caused by particles that turn into the hole and ballistically impact the leeward wall. These particles are likely dominated by inertia and penetrate the boundary layer too rapidly for the interaction with small scale turbulent structures to have an effect on the trajectory. This gives additional justification for using the more economical DRW model.

$$u' = v' = w' = \sqrt{\frac{2k}{3}} \qquad (13.1)$$

If particles exit one of the two outlets in the domain, the particle tracking is terminated. When a particle impacts a wall, the OSU Deposition Model is used to predict whether that impact will result in a stick or a rebound. In the case of a rebound, the model dictates the rebound velocities to the DPM model, and the particle is reinjected into the domain with the prescribed velocity vector. The OSU Deposition Model treats the particle as a cylinder impacting end on and calculates the elastic-plastic energy exchange of the inbound kinetic energy. The primary mechanical properties that must be carefully chosen and provided to result in accurate predictions are the particle yield stress, a composite elastic modulus of the particle and the surface material, and the surface free energy to approximate the adhesive forces at contact.

As mentioned, the predictive capability of the model hinges on reliable mechanical properties. Whitaker et al. [27] used a Particle Shadow Velocimetry technique to amass a large database of particle rebound data over a wide range of impact velocities, angles, and temperatures for different dust types and their individual constituents. The authors reported comprehensive mechanical property data that when implemented in the model replicated the mean behavior seen in the PSV rebound data. The properties for ARD at 950 K impacting an Inconel surface are the closest match to the experiments in this study and were implemented with the OSU Deposition Model. A list of the ARD mechanical properties used is provided in Table 13.1. Of special note is the correlation for determining the yield stress (σ_y). Whitaker et al. proposed a velocity dependent yield stress that captures rate of strain effects that occur when particles impact at higher velocities. This concept and the trend shown by Whitaker et al. was reconstructed using the Johnson-Cook [45] yield stress relationship as adapted by Plewacki et al. [46]. The Johnson-Cook model allows for an empirical handling of both rate of strain effects in addition to particle softening effects that

Table 13.1: Particle mechanical properties used with the OSU Deposition Model for ARD.

Mechanical Property Name	Value
Particle density (ρ_p)	$2727 \; {}^{kg}/_{m^3}$
Particle specific Heat (c_{pp})	$714 \; {}^{J}/_{kg-K}$
Yield stress (σ_y)	$if \; \left[100\left(1 + 1.8\log\dfrac{U_p}{27.86}\right) < 100\right] : 100 \; MPa$ $else$ $100\left(1 + 1.8\log\dfrac{U_p}{27.86}\right) MPa$
Composite elastic modulus (E_c)	$72.3x10^9 \; GPa$
Surface free energy (γ)	$0.8 \; {}^{N}/_{m}$
Dynamic friction coefficient	0.6

occur for particles at higher temperatures than seen in these experiments.

Mesh morphing is becoming a common technique used to model the coupled nature of the building deposit and the influence on the flow-field and resulting particle trajectories. Bowen et al. [47] used the technique to simulate the growth of a cone shaped deposit in an oversized impinging jet flow. Figure 13.1 shows the resulting deposits the authors built computationally using a multi-step morphing approach and a single-shot approach. The single-shot approach predicted the sticking on the initial clean surface and then assumed all of the mass delivered in the experiment built in one-step. This completely ignored the effect that the deposit has on the flow, particle trajectories, and sticking predictions as it develops. The resulting structure (right) is seen to be quite unrealistic compared to the experimental deposit (left) and creates a plateau rather than a smooth cone. Using the multi-step mesh morphing progression, the resulting deposit (middle) shows a more reasonable resemblance to the experimental cone. This work sensitized the need to deliver smaller amounts of dust to incrementally move the mesh boundaries and allow for recalculation of the flow-field and particle trajectories.

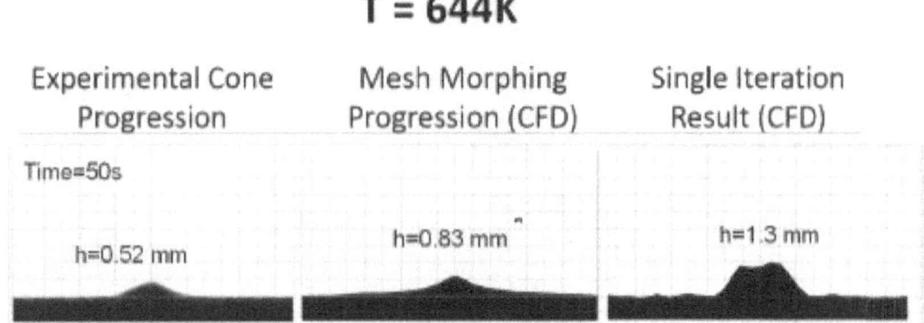

Figure 13.1: Bowen et al. [47] comparison between experimental impingement deposit formation (left), multi-step mesh morphing structure (middle), and single iteration morphing structure (right).

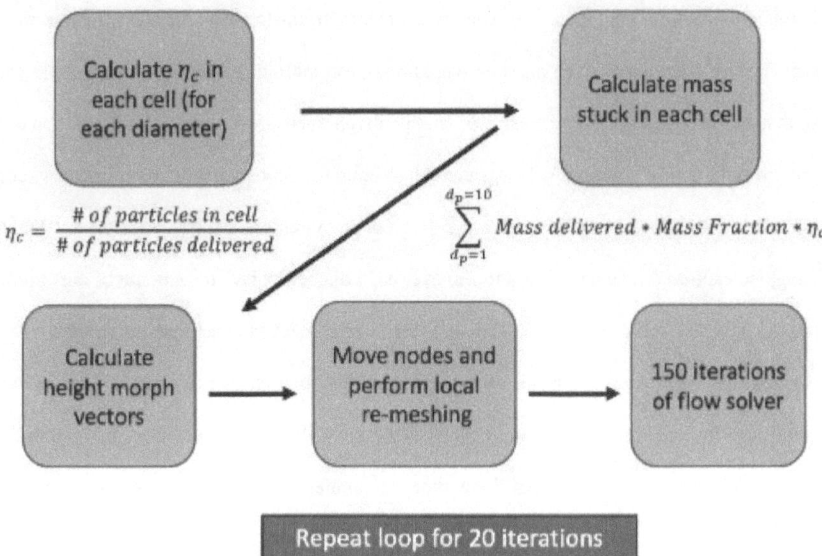

$$\eta_c = \frac{\#\ of\ particles\ in\ cell}{\#\ of\ particles\ delivered}$$

$$\sum_{d_p=1}^{d_p=10} Mass\ delivered * Mass\ Fraction * \eta_c$$

Figure 13.2: Mesh morphing routine flowchart

A flowchart of the mesh morphing routine used to model the effusion hole deposit growth is provided in Figure 13.2. After all of the particles have finished tracking, the first step is to calculate capture efficiency in each wall cell. This is calculated simply by dividing the number of particles predicted to stick in each cell by the 10,000 particles delivered for each diameter.

This number capture efficiency in each cell is then converted to a mass capture efficiency. Each diameter studied represents a fraction of the total mass delivered. This fraction is determined using the cumulative mass distribution data provided by PTI for the 0-10 μm ARD. The mass fractions for the 10 sizes studied all sum to 1. The mass weighted capture efficiency for each size is found by multiplying the number capture efficiency from step 1 by the mass fraction for each size. The total amount of dust mass delivered per mesh morphing iteration was carefully chosen so that the displacement of the mesh was small enough to ensure accurate coupling of the deposit growth and the flow solution, but large enough to build the deposit without requiring an excessive number of iterations. 7.5×10^{-9} kg was delivered per iteration. Multiplying this by the mass weighted capture efficiency and summing over each diameter gives the amount of dust predicted to stick in every cell per iteration. The next step is to convert the dust mass to volume and from that calculate the displacement for each node to achieve that change in volume. The conversion from mass to node displacement used by Forsyth et al. [48] was implemented, but without some of the additional complexities that the authors recommended. Figure 13.3 provides a view of a tetrahedral mesh to help visualize how each node displacement is determined. A user-defined function (UDF) loops through each node on boundary mesh and calculates the displacement vector using the dust mass in the faces adjacent to that node. In Figure 13.3, the grey circle highlights the node of interest and the faces that border the node are highlighted in red. The mass in each triangular face is evenly split among its three nodes. The total mass contributing to the highlighted node (m_n) is then the sum of the mass in the surrounding faces (m_f) divided by three as shown by Equation 13.2. The mass is then converted to a volume by dividing by the density of ARD as well as the packing factor (PF) of ARD as shown in Equation 13.3. The PF accounts for porosity of the deposit and is defined as the volume of dust in the deposit by the total volume of structure including the air gaps between particles as defined by Equation 13.4. Bowen et al. [47] showed that the PF of ARD in an impinging jet deposit structure remained approximately constant with temperature and velocity at an average value of 0.28. This means that deposit structure is about 72% air by

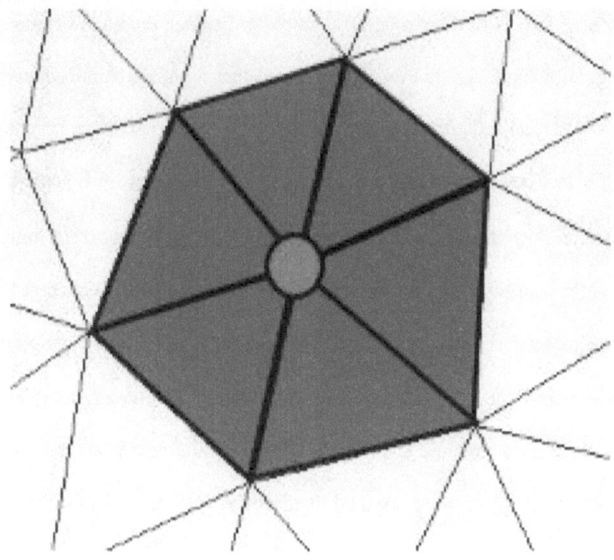

Figure 13.3: Tetrahedral surface mesh with central node (grey) and adjacent faces (red).

$$m_n = \sum_{f_{adj}} \frac{m_f}{3} \qquad (13.2)$$

$$\forall_{deposit} = \frac{m_n}{\rho_p * PF} \qquad (13.3)$$

$$PF = \frac{\forall_p}{\forall_{deposit}} = \frac{\rho_{dep}}{\rho_p} \qquad (13.4)$$

volume. The next step is to determine how much to displace the node of interest to create the desired change in volume. To do this, it is assumed that the node is moved away from the wall along its surface normal and creates a pyramid. The area of the base of the pyramid (A_b) is simply the sum of the face areas adjacent to the central node as given by Equation 13.5. The node displacement vector ($\vec{h}$) then is calculated through Equation 13.6 using the formula for the volume of a right pyramid. The node normal unit vector ($\widehat{n_n}$) is taken as the area averaged face normal unit vector of the adjacent faces (n_f) as defined by Equation 13.7. This calculation is made for each boundary node. Forsyth et al. emphasized that when using this approach, there is inherent overlap in the projected pyramids because each face is shared by 3 nodes. When the nodes are moved to their new locations, this can cause the total displaced volume to be larger than the intended. To overcome this Forsyth et al. used an iterative procedure to make slight adjustments each node location to give the desired change in volume. In this study, it was found that the error in the volume displacement increased with each iteration, but never exceeded 3% of the target volume. It was determined that this error would have a very minimal impact on the results and conclusions made, so an iterative routine to correct that error was not implemented in the interest of faster computation time.

$$A_b = \sum_{f_{adj}} A_f \qquad (13.5)$$

$$\vec{h} = \frac{3\forall_{deposit}}{A_b} * \widehat{n_n} \tag{13.6}$$

$$\widehat{n_n} = \frac{1}{A_b} \sum_{f_{adj}} \widehat{n_f} \tag{13.7}$$

The dynamic meshing feature native to Fluent is used to morph the mesh. Another UDF communicates the new position vectors $(\overrightarrow{x_f})$ for the boundary nodes to the dynamic meshing tool. $\overrightarrow{x_f}$ is calculated by summing the current position vector of the node $(\overrightarrow{x_o})$ and the displacement vector $(\vec{h})$. The dynamic meshing capability in Fluent has many useful features including the ability to re-mesh local cells by defining a cell skewness threshold. This skewness was evaluated every time the mesh was morphed, and re-meshing was performed for cells with a skewness exceeding 0.9. Another feature is a smoothing function that controls how the nodes in the internal volume shift when the wall boundary nodes are displaced. A spring based smoothing method is employed, where the edges of cells are treated as a network of interconnected springs with a given spring constant. When subject to a "force" that is the wall displacement in this case, the internal cell nodes move according to the spring constant assigned by the user. A low spring constant of 0.1 was used which allowed the internal mesh to deform with relative ease without causing cells to intersect and create an invalid mesh. The benefit of the built-in re-meshing and smoothing capabilities in Fluent is that it allows for the maintenance of a high-quality grid, without requiring the entire domain to be completely re-meshed. This results in a very efficient morphing action and requires no exporting and re-importing of the geometry which can cause inherent truncation of geometric features in other commercial CFD packages [49].

The final step after the mesh has been updated is to re-calculate the flow-field. The deposit growth is simulated gradually enough that the changes in the flow-field are relatively minor, and thus only 150 iterations are required to re-converge from the prior solution. The percent reduction in the mass flow through the hole (MFR) is determined and the iteration concludes. These steps are carried out for 20 iterations, which is sufficient to build a deposit representative of what is seen in the effusion holes after the experiments.

Chapter 14. Mesh Morphing Results

The final deposit structure after 20 iterations of mesh morphing at 1 atm using the spherical drag correlation and 0-10 μm ARD is shown in Figure 14.1. An isometric and a top-down view are provided, and a deposit thickness contour is shown on the surface. The outline of the clean effusion hole is shown with a dotted red line in each image for reference. The morphed structure causes approximately 3.5% MFR. To compare this to the experimental deposit, another experiment was run at 1 atm and the dust delivery was stopped after approximately 5% MFR was reached. A top-down microscope image of the resulting deposit is provided in the bottom-right section of Figure 14.1. The morphed deposit shows close resemblance in terms of shape to the experimental deposit. The tongue that causes blockage is prominent in both. Additionally, the experimental deposit has a small "skirt" of dust that builds on the plate behind the leeward side of the hole. This is where the shell structure begins to develop as more ballistic particles impact and settle in the area. The computations also predict this structure quite well with regards to the general thickness and shape.

Figure 14.2 gives the same comparison, but now for the morphed deposit when the 20 iterations are performed using the non-spherical drag model with $\phi = 0.65$. Generally, the structure looks nearly identical to what was built with the spherical drag model. There is a slight reduction in the overall deposit thickness in the tongue structure, but the changes are nearly imperceptible. This is not unexpected when considering the discussion of the validity of the spherical particle assumption in Chapter 10. The range of Re_p for 0-10 μm at 1 atm for the diameters tracked is approximately

0-25. The smaller particles that dictate the sticking (0-3.5 μm) however have a smaller Re_p of approximately 0-5. At these lower Re_p, Figure 10.3 shows that the drag coefficient curves all collapse to nearly the same value no matter what the specific value of ϕ is. Thus, for this experiment at 1 atm, the spherical particle assumption is a suitable one and allows for the prediction of a realistic structure.

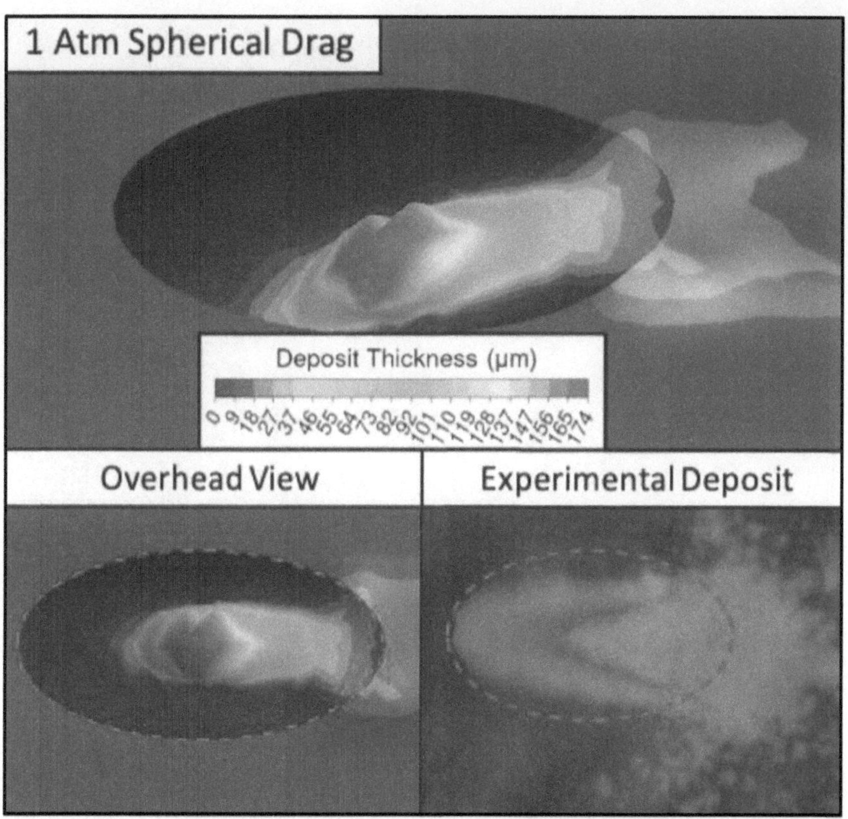

Figure 14.1: 1 atm mesh morphing final deposit with deposit thickness contour (top, bottom left) when the spherical drag model is used with 0-10 µm ARD. Experimental deposit microscope image at approximately the same level of blockage as the CFD for comparison (bottom right).

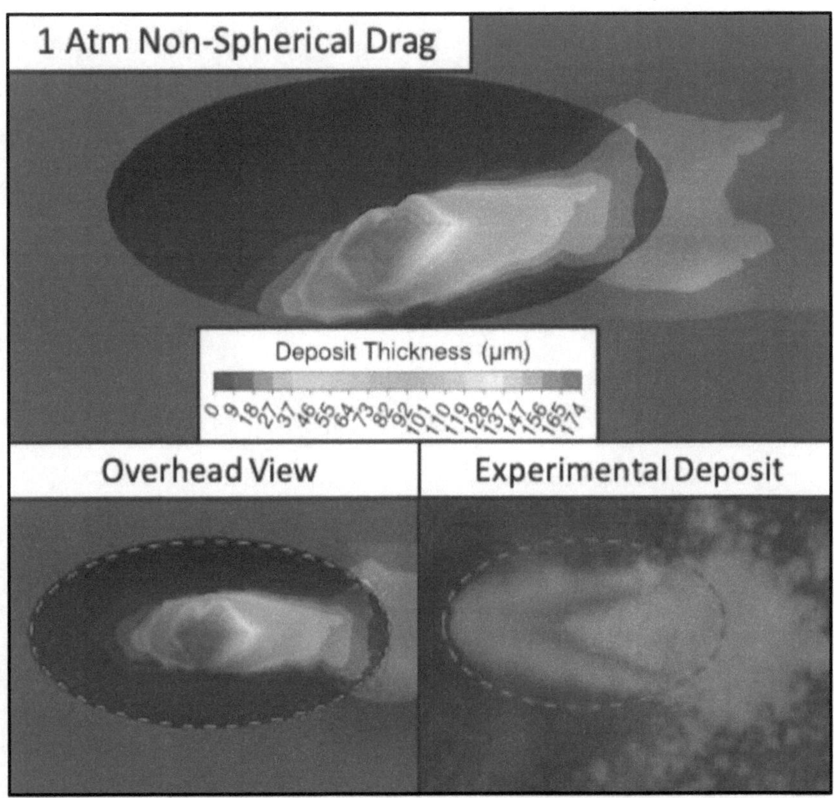

Figure 14.2: 1 atm mesh morphing final deposit with deposit thickness contour (top, bottom left) when the non-spherical drag model ($\phi = 0.65$) is used with 0-10 µm ARD. Experimental deposit microscope image at approximately the same level of blockage as the CFD for comparison (bottom right).

Figure 14.3 shows the final mesh morphing result using the spherical drag correlation at 15.77 atm. The first important difference to notice when comparing to the 1 atm deposits is the limit of the deposit thickness contour. The maximum deposit thickness is only 71 μm as opposed to 174 μm at 1 atm. The sticking has clearly decreased, and the shape again shows generally good agreement with the experimental deposit. The primary difference is the reduction in the tongue structure that dictates the blocking. Particles impacting in this region are doing so at reduced impact angles and are able to roll from the surface, reducing the sticking in this area. The added drag at high pressure enables this phenomenon, and it is captured by the particle tracking in combination with the OSU Deposition Model without changing any inputs to the model. The deposit builds up primarily along the leeward side of the hole and on the effusion plate behind the leeward edge of the hole. If further iterations were performed, one could reasonably expect that the shell behind the hole would grow over the hole as it was shown to do in the experiment at high pressure. Although the spherical drag model captures a general reduction in the capture efficiency of the particles as pressure increases, there is a noticeable amount of localized growth in the tongue region. While less pronounced and developed than at 1 atm, it is still a contributor to the blockage and is an unrealistic feature that is not seen in the experiment at 15.77 atm.

The use of the non-spherical drag model is found to greatly improve the prediction by eliminating the sticking in the tongue region. The range of Re_p for 0-10 μm ARD at 15.77 atm is approximately 0-200. The increased flow density and resulting increase in Re_p cause the spherical drag model to underpredict the total particle drag. Figure 10.3 shows a significant separation between the spherical and non-spherical drag coefficient curves at these higher Re_p. The use of the non-spherical drag correlation causes the smaller particles in the distribution to impact at an even lower frequency than is predicted using the spherical model. The particles that do impact do so at even smaller angles and rarely stick in the tongue region. The deposit build-up is limited to the leeward side of the hole and shows striking resemblance to the experimental deposit.

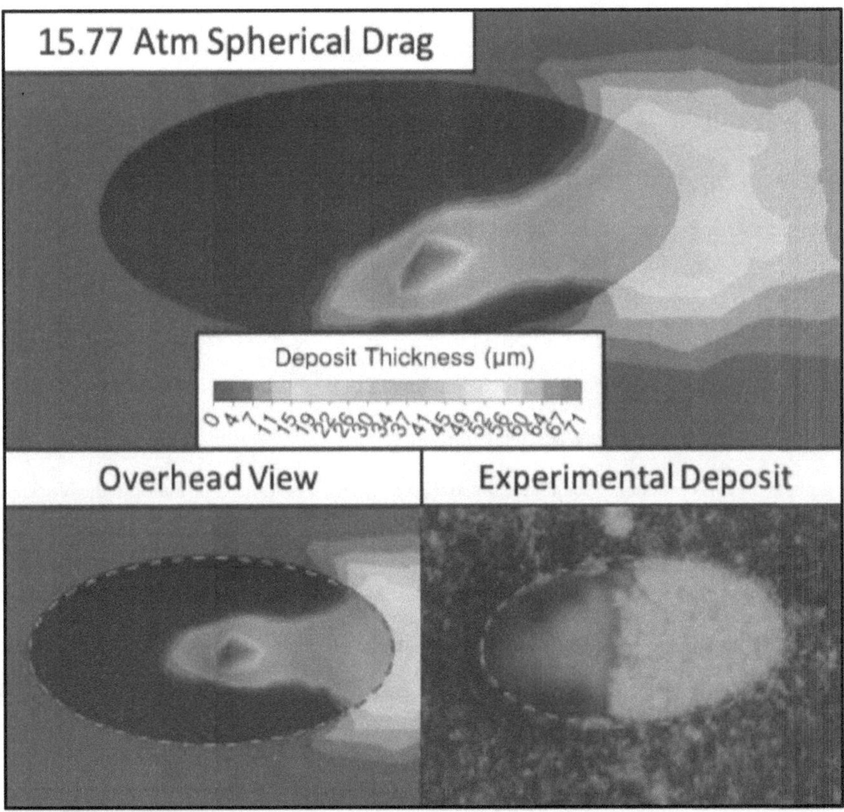

Figure 14.3: 15.77 atm spherical mesh morphing final deposit with deposit thickness contour (top, bottom left). Final experimental deposit microscope image for comparison to CFD (bottom right).

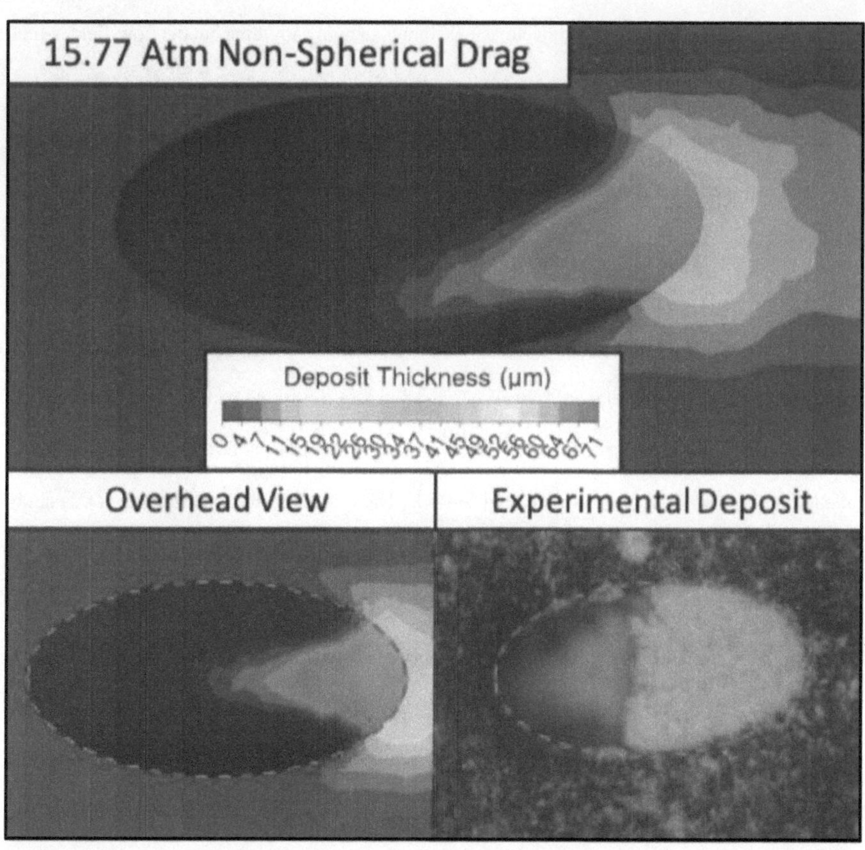

Figure 14.4: 15.77 atm non-spherical ($\phi = 0.65$) mesh morphing final deposit with deposit thickness contour (top, bottom left). Final experimental deposit microscope image for comparison to CFD (bottom right).

The examination of the final mesh morphing structures provides strong qualitative evidence that the decreases in blockage and particle deposition with pressure are caused by increased particle drag and the effect on the particle trajectories. Without changing anything but the flow-field, which has a much higher fluid density, the deposit shapes are well-predicted by the model. It is important to remember, however, that no erosion physics were implemented in the modeling. Libertowski et al. [49] developed the first ability to include effects of erosion with the OSU Deposition Model, however the results of that effort were very preliminary. The study developed empirical erosion correlations for 40-80 μm ARD eroding an existing over-sized impingement cone deposit. These correlations provide a diameter scaling for the erosion; however, they appear to be specific to the experiments they were developed from. When used in this flow-field and for 0-10 μm particles, the amount of mass that is predicted to erode in a given iteration is three orders of magnitude smaller than the amount of mass predicted to stick. The erosion in the experiments in this study was proven to be a substantial factor, and thus a more robust and generalized erosion model would be necessary to capture those physics.

Although the structures show general resemblance, no consideration has yet been given to the amount of dust delivered in the morphing procedure, and how the MFR vs grams delivered compares to the experiment. Doing this analysis provides insight into the effect of increasing drag on the smaller particles compared to the effect of drag on large particles that cause erosion. The first effect is captured in the modeling, while the second is of course neglected. The MFR per iteration of mesh morphing is easily calculated in the CFD for each of the four simulations. To compare the single hole CFD results to the experiment, the amount of mass in the CFD must be scaled to match the 357-hole experimental effusion plate. This conversion was done by realizing that all of the mass delivered in the experiment either stuck to the plate or exited the effusion holes. The amount of injected mass (7.5×10^{-9} kg) that either stuck to the plate or exited the effusion hole was tracked for each iteration and scaled by 357. This makes the assumption that an equal amount of dust enters or sticks in each hole. In the experiments, an average of 10% of the delivered dust collects at the end of the diffuser. This region is not modeled with the single effusion hole geometry, and thus it is appropriate to account for this by dividing the computationally scaled mass by 0.9.

The MFR vs cumulative grams smaller than 3.5 µm delivered for all four morphing simulations are plotted against the corresponding experimental results. Considering only the mass delivered that is smaller than 3.5 µm is necessary for purposes of comparison between the two distributions for the reasons described earlier in the discussion of using $BPG_{<3.5}$ as a metric for comparing rates of blockage. The top portion of the Figure 14.5 plots the 1 atm results. The spherical (red) and non-spherical (green) MFR trends overlap nearly perfectly, with only a slightly reduced slope for the non-spherical simulation. This overlap is consistent with the finding that the final morphed structures were also nearly identical. This provides confirmation that the spherical assumption is valid for modeling purposes at 1 atm. What is more illuminating is the fact that the computational trends closely follow the 0-3.5 µm experimental MFR rather than the 0-10 µm trend. Although the 0-10 distribution is modeled, the lack of an erosion model causes the structure to build more quickly in the computations than it does in reality. The larger particles in the simulations essentially do not participate, so when the MFR is normalized by the smaller particles that are delivered, the trend is found to match the 0-3.5 experiment. This shows that the model does an accurate job capturing the sticking and subsequent blocking of the effusion hole by the small particles in the distribution. This also confirms that normalizing MFR by the particle sizes beneath the sticking threshold ($\cong 3.5\ \mu$m) causes curves from different dust distributions to collapse as Wolff et al. suggested.

In the bottom plot in Figure 14.5 for 15.77 atm, the computational MFR trends again appear to follow the 0-3.5 μm experimental curve rather than the 0-10 μm. There is more noticeable separation between the spherical (red) and non-spherical (green) curves at this pressure which is expected because of the more influential effect of sphericity on the drag prediction at higher Re_p. Neither drag correlation however causes the MFR to shift to zero as is seen in the 0-10 μm experiment. This again is due to the lack of erosive physics, which in the experiment removes any build-up in the tongue region. Comparing the 1 atm and 15.77 atm MFR trends for the 0-3.5 μm ARD, it is seen that the slope is approximately a factor of three lower at 15.77 atm. This reduction with pressure is caused by the effect of drag reducing the number and the way the small particles impact. The model captures this physics with increased pressure. The 0-10 μm experimental MFR shows a factor of reduction of more than two orders of magnitude, although it is hardly evident from the figure due to the very minimal amount of blockage for the 0-10 μm at 15.77 atm. This lends credence to the theory that the particles below 3.5 μm have the tendency to stick less aptly at higher pressures, while the erosion by particles larger than 5 μm is significantly enhanced as pressure increases. A high-fidelity erosion model is clearly needed not only to accurately replicate the rate of deposition for dust distributions where erosion is non-trivial, but also to capture the way that increasing the fluid pressure and density augments the erosion.

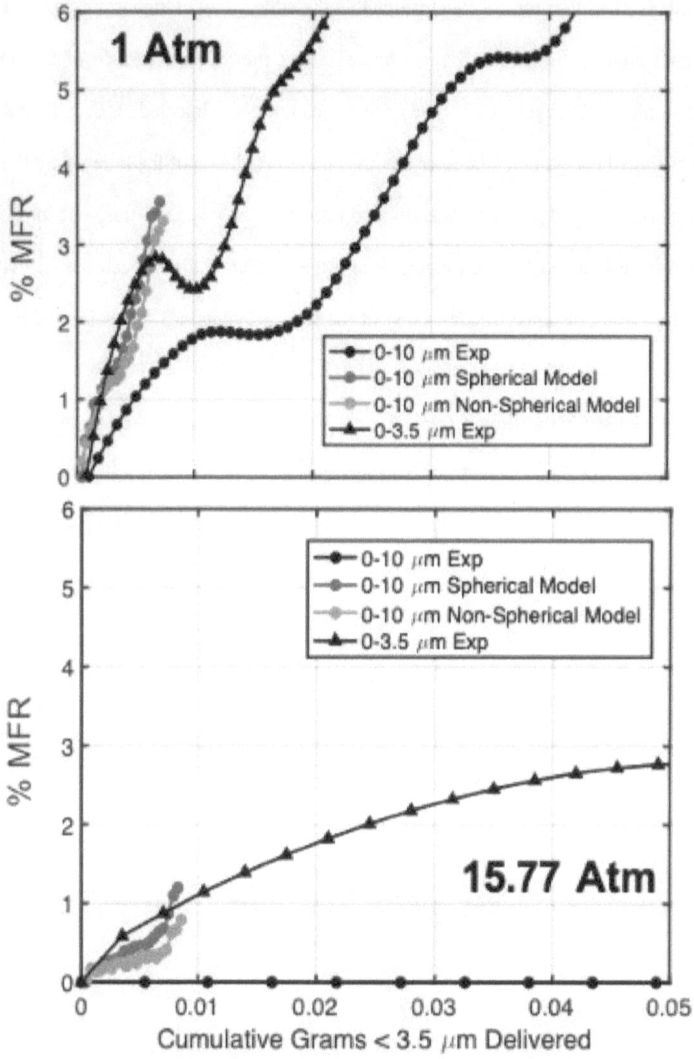

Figure 14.5: Mesh morphing MFR versus grams delivered smaller than 3.5 μm for spherical (red) and non-spherical (green) drag. Experimental MFR for each size distribution studied plotted in black.

To provide justification for the conclusions that the added drag at higher pressure reduces the particle impact angles, the median impact angles along the leeward surface of the hole were calculated and plotted in Figure 14.7. Five of the sizes that were injected (1, 2, 3, 5, and 10 µm) are plotted to show how particle diameter ties in. The leeward side of the hole where the deposit structures are seen to build primarily was divided into ten evenly spaced bins. The impact angles in each bin were calculated for each diameter and the median angle in each bin is plotted as a function of the wetted distance fraction (x/S) of the leeward edge. x/S = 0 is located at the inlet plane of the hole, while x/S = 1 is located at the exit plane as shown in Figure 14.6. Figure 14.7 shows that the median impact angle for 1, 2, and 3 µm particles decreases by approximately a factor of two from 1 to 15.77 atm. 5 and 10 µm particles have a similar reduction although not quite as significant. Also of note is that the non-spherical (NS) impact angles are lower than the spherical (Sph) by about 5-10 degrees.

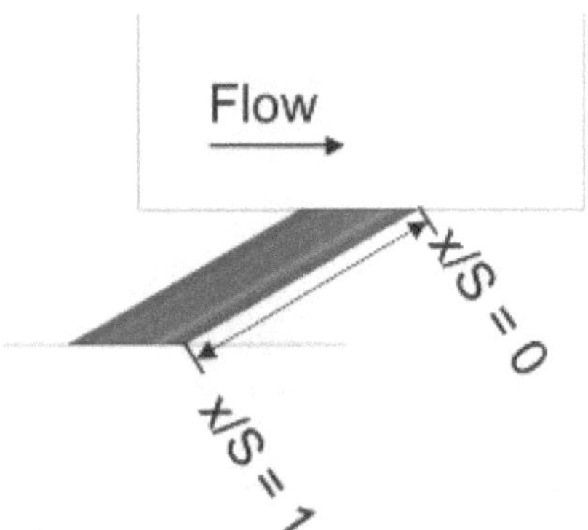

Figure 14.6: Schematic of the effusion hole with x/S defined for Figure 14.7.

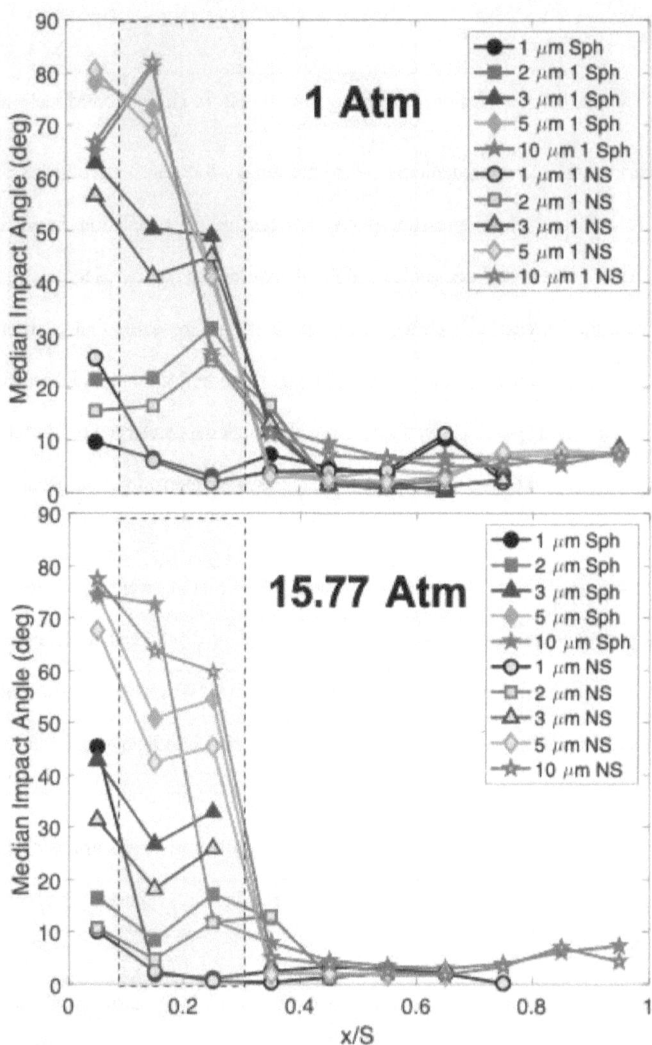

Figure 14.7: Median particle impact angle vs x/S on leeward (impact) side of the clean effusion hole for 5 d_p at 1 atm (top) and 15.77 atm (bottom). Transparent yellow box depicts region where the tongue structure typically builds.

Chapter 15. Conclusions Regarding High-Pressure Deposition Modeling

The experimental and computational effusion cooling study presented further sensitizes the first order role that absolute pressure plays in determining deposition in internal cooling geometries. An order of magnitude analysis of the shear moment compared to the counter-moment created by adhesion forces between a particle and a surface revealed that even at the highest pressures seen in a modern engine, the shear moment is an order of magnitude less than adhesion. This does not rule out the potential for shear removal of bulk deposit structures, but it does suggest that increasing shear forces are not driving the decreases in deposition that occur as pressure is increased.

The discharge coefficient effects, as suggested by Sacco et al., were also investigated. While it was found both experimentally and computationally that the discharge coefficient increases with absolute pressure in these experiments, a targeted experiment was used to show that this plays a secondary role in deposition. The average flow velocity increases by approximately 22% from 1 atm to 15.77 atm despite fixing the pressure ratio and flow temperature. By reducing the flow velocity at 15.77 atm, it was shown that the mass flow reduction versus grams of dust delivered trend was practically unchanged. This left the third mechanism suggested by Sacco et al. as the likely candidate: drag effects.

The increasing drag on the particles that occurs with higher pressure was shown to be the dominant mechanism causing reduction in deposition and coolant hole blockage in the experiments. The effective Stokes number and its importance were discussed, and it was theorized that the reduction in the effective Stokes number due to increased particle drag at higher pressures allows particles to follow streamlines more easily. This leads to fewer impacts with the surface as well as low angle impacts that have an augmented ability to roll and detach from the surface. The role of erosion by particles larger than 5 µm was demonstrated in another targeted experiment. It was shown that 0-10 µm ARD at 15.77 atm could erode the deposit in an effusion hole deposit and cause a mass flow recovery. Comparing a smaller distribution (0-3.5 µm ARD) to a larger one (0-10 µm) by using a modified blockage per gram metric, it was hypothesized that the increased drag causes 0-3.5 µm particles to stick less frequently and allows particles larger than 5 µm to erode more aggressively.

A computational investigation using a mesh morphing routine to simulate the deposit build up in the flow-field was performed. The results provided further evidence to confirm the hypothesis about elevated drag with increasing pressure. The simulations were performed using spherical and non-spherical particle drag correlations and emphasized the importance of using non-spherical models when tracking particles at elevated particle Reynolds numbers. Failure to do so would result in underestimated drag and overestimated deposition rates. These results indicate that high-pressure effects are substantial in deposition experiments and modeling. To capture the proper physics with respect to small particles with the potential to stick and build on a surface, the flow density must be matched, and an accurate non-spherical tracking scheme must be employed. For distributions with particles larger than 5 µm, erosion is likely to occur and could be more pronounced at higher flow pressures. Striving for a higher fidelity erosion model that can capture this trend should be an emphasis moving forward.

Chapter 16. Extending Mesh Morphing Predictive Abilities

In the mesh morphing procedure of the previous study there is no consideration given to the impact that the deposit structure has on the thermal performance of the cooling scheme. As particles build on metal surfaces in an internal cooling section of an engine, the cooling effectiveness can vary for two primary reasons. First, blockages can occur that reduce the mass flow rate of the coolant. This can reduce the heat transfer into the coolant, and in the case of film cooling can cause the film to develop in a non-optimal fashion. Both would likely result in an increased metal temperature. The second impact is that the particulate layers cause a layer of thermal insulation on the metal surfaces. When deposit forms on the outer surface of a turbine vane or blade, the deposit creates an extra layer of thermal resistance that can cool the metal surface. This benefit is typically outweighed by the aerodynamic impact, however. In internal cooling channels, one would expect an insulating layer of dust to prevent the effective transfer of heat from the metal to the coolant, causing a decrease in overall cooling effectiveness and an undesirable increase in metal temperature.

To date, the majority of mesh morphing efforts in the open literature for deposition prediction have only considered the impact on the fluid flow-field. Forsyth et al. [48] developed the technique to mesh the fluid boundaries of a tetrahedral mesh in ANSYS Fluent as described in Chapter 13. The authors used the procedure to model an impingement cooling array as well as an effusion cooling hole subjected to deposition and were able to replicate the general shapes reasonably well. Bowen et al. [15] and Libertowski et al. [49] implemented the same framework in Star-CCM+ and morphed the buildup and erosion respectively of an oversized impingement cone using a polyhedral mesh. None of the aforementioned studies however made any consideration for the heat transfer through the deposit structure. Paul et al. [50] used a high-fidelity in-house Large Eddy Simulation (LES) solver to model the four-way coupling of the momentum and energy exchange between the particle-to-particle and particle-to-fluid interactions in an impinging jet. After predicting deposit on the surface, the deposit cells were modeled as a porous media and the heat transfer through the structure was predicted. García Pérez et al. [51] modeled ash deposition on a 4-tube boiler bank in ANSYS Fluent on a 2-D mesh. The authors created an initial thin layer of solid cells on the tube surfaces and after predicting deposition morphed that thin layer of solid into the adjacent fluid. The solid region in this case was also modeled as a porous media and the heat transfer was calculated through the deposit. Both of these studies added a degree of complexity in calculating the heat transfer through the deposit and considering the effects of porosity, however, neither case modeled the heat transfer through the solid metal substrate.

When the fluid domain is modeled without a solid domain, the thermal boundary conditions must be enforced at the walls to solve the energy equation. This is usually done with either an isothermal (constant temperature) or constant heat-flux condition. Neither of these prescribed physics is remotely valid under real engine operation, but both are usually easy to estimate or measure and can allow for sufficiently accurate modeling of the fluid domain. When the deposit growth is modeled by morphing the mesh, the effect on the flow in the internal fluid cells is predicted, but the thermal conditions on the walls remain unchanged. This is extremely limiting in the sense of using deposition modeling to predict how the metal temperature is altered.

To truly make mesh morphing a viable tool for making thermal predictions, it must be implemented in conjugate simulations. Conjugate simulations model the solid regions as well as the fluid. The heat transfer through the solid is calculated by solving a reduced form of the energy equation on a meshed representation of the solid geometry. Rather than enforcing thermal boundary conditions on the external surfaces of the solid, they are inferred from the adjacent fluid cells by what is called a contact interface. The solid temperature is coupled to the fluid domain, and as such changes in the fluid due to deposit growth would necessarily influence the prediction of the metal temperature. The ability to morph both the fluid and solid boundaries in a coordinated fashion that preserves the solid and fluid interfaces is only the first step to achieving this coupling. The second step is to identify cells in the solid region that represent deposit and thus have thermal and mechanical characteristics that affect both the heat transfer and the deposition predictions respectively. Porous dust can have thermal conductivity values approximately 2 order magnitudes lower than that of typical high-temperature nickel-alloys, so assigning the cell conductivities correctly is paramount to achieving accurate thermal predictions.

The objective of the following study is to develop a conjugate mesh morphing ability that can achieve both of those requirements. To test the predictive capabilities of the conjugate deposition framework, an oversized (roughly an order of magnitude larger than typical engine impingement holes) jet impingement deposition experiment is run with thermal imaging and synchronized high-speed video of the deposit evolution. The model will attempt to replicate both the particle accumulation and the ensuing thermal impact.

Chapter 17. Experimental Setup and Procedure

Deposition in an impingement cooling scheme was replicated in a low-pressure deposition facility at the OSU Aerospace Research Center. A full facility schematic is provided in Figure 17.1. Impinging jets as mentioned previously are commonly used for internal cooling in turbine and combustor geometries because they provide high heat transfer coefficients with minimal geometric complexity. When dust contaminates the cooling air in an impinging jet, a distinct and often axisymmetric cone shaped deposit builds on the target wall, insulating the wall and limiting the heat transfer from the hot wall to the coolant. This causes thermal degradation as well as other performance related issues that make the scenario a prime candidate for deposition studies.

Numerous authors, including Sacco et al. [22] and Clum et al. [52] have detailed the effects of deposition in impingement geometries at the engine scale. Bowen et al. [15] and Libertowski et al. [49] both used an oversized impingement hole to build an enlarged deposit cone. While not specifically relevant to jet engines, the authors opted for an impingement hole approximately an order of magnitude larger (~6.35 mm) than what is typically used in actual engine hardware (~0.6 mm) because it allowed for enhanced visual access and more accurate measurement capabilities. The authors noted that while the deposit structures that formed were obviously larger than in practice, they form in a geometrically similar fashion. As long as the hole length to diameter (l/d) and gap to diameter (z/d) are fixed, the cone dimensions and volume can be normalized and show good agreement with results of smaller geometries. For the purposes of informing new deposition

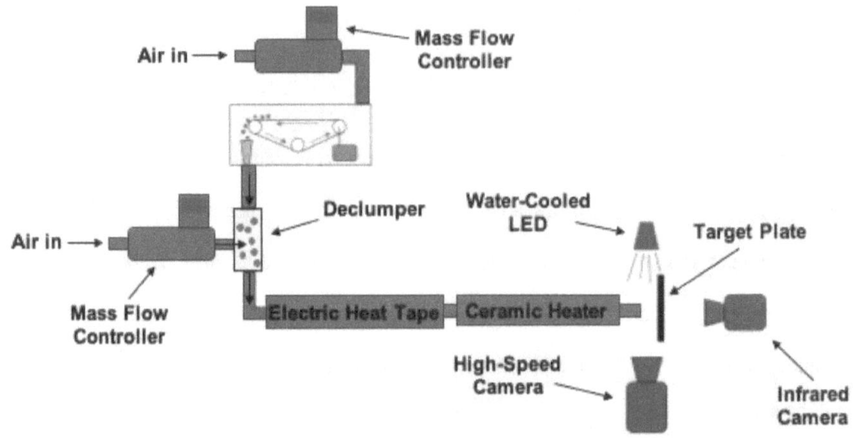

Figure 17.1: Impinging jet deposition facility schematic.

physics and enhancing the computational modeling abilities, these are convenient and maintain relevance.

1.4 grams (± .0003 g) of 0-5 ARD is baked to remove moisture, weighed, and finally loaded evenly onto a 12" long conveyor belt. This smaller size distribution of dust was chosen because it results in a fairly large deposit structure without requiring an excessive amount of dust to be delivered. The cumulative mass distribution provided by PTI is plotted in Figure 17.2, and the 12 discrete diameters injected in the latter modeling section are marked. The box housing the conveyor belt is sealed, and a 0-0.00493 kg/s mass flow controller is set to feed 0.000373 kg/s (± .0000014 kg/s) of low-pressure air into the conveyor box. When the belt is started and particles fall into the funnel and through the drop tube, this air carries the falling particles into a declumping

114

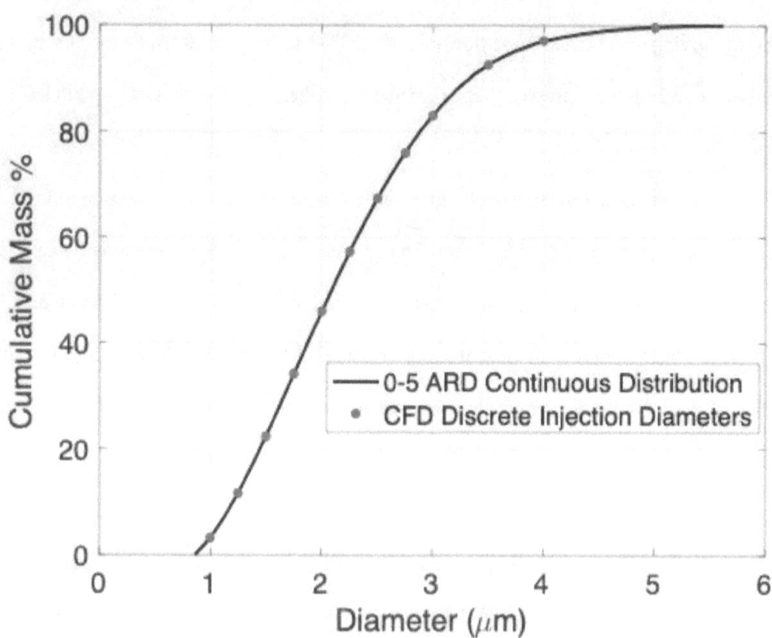

Figure 17.2: 0-5 ARD dust cumulative mass distribution vs particle diameter (black). CFD discrete injection diameters plotted along the distribution (red circles).

section. A common issue with micron sized particles is the agglomeration of small particles into larger clumps. This presents a challenge in modeling the particles, as the measured size distribution is altered entirely. This makes modeling the drag and impact mechanics a more arduous task, one that is undesirable for most fundamental deposition experiments. The declumping is accomplished with a secondary stream of air provided by a second 0-0.00197 kg/s mass flow controller set to 0.00033 kg/s (± .0000007 kg/s). This stream of air mixes with the primary air source in the drop tube as shown, giving a combined mass flow rate of 0.00071 kg/s (± .0000016 kg/s). The mixing causes turbulent and viscous stresses that overcome the cohesive forces between particles and break them apart.

After the two streams mix out, the declumped particle laden air enters a 10' section of Inconel tubing with an inner diameter (d) of 6.35 mm. The upstream section of the tubing is wrapped in heat tape, while the downstream section closest to the tube exit is heated by a Watlow 850 W ceramic heater. These combine to heat the centerline of the tube exit flow to 894 K. This combination of mass flow and average flow temperature results in an average exit flow velocity of approximately 56.5 m/s. A 76.2x76.2x3.175 mm Hastelloy X target plate is secured at a distance of 2 tube diameters (z/d = 2) downstream of the tube exit.

A Photron Fastcam SA-Z high-speed camera with a Nikon ED 200 mm lens is focused so that it is directly aligned with the frontside of the target plate to record a video of the deposit growth during the test. A 620 nm water-cooled LED light is placed directly opposite the camera, allowing the profile of the building deposit to be clearly visualized from the projected shadow. This technique is known as Particle Shadow Velocimetry (PSV) and is often used to visualize airborne particles and record their trajectories. The resolution of the camera with the aforementioned lens extension is approximately 20 μm, and thus none of the particles in this distribution are visible as they exit the pipe and impact the target. The absence of visible particles provides confidence that large clumps are being successfully separated in the declumper section. The focus is adjusted so that the centerline of the pipe is in focus, assuring that the center of the developing cone structure is also in focus.

A Cedip Silver Infrared (IR) camera is placed behind the target plate and is used to extract the temperature field on the backside of the target plate. The integration time of the camera is set to 25 μs. To calibrate the IR camera, a 1 mm diameter stainless steel k-type thermocouple is spot welded to the backside of the target plate such that the tip of the thermocouple is centered on the plate. As the flow is gradually heated, the plate temperature increases, and IR snapshots are captured with every 45 K increase in the thermocouple reading. 18 calibration images are taken over a range of 394 to 630 K. In post-processing, a three-point average of the digital level around the thermocouple tip is acquired and used to generate a temperature versus digital level curve. A logarithmic trend with an r-squared value of 0.9987 is fit to the curve and used to convert the thermal images during the deposition event to temperature contours.

Once the calibration is complete and steady flow conditions are reached, the plate is backed away from the tube and allowed to cool before it is removed. A new plate that is cut from the same metal stock and with equal dimensions replaces it, which allows for clean IR images with no interference from the thermocouple. The plate is moved toward the tube exit until it is spaced at the correct distance and a waiting period occurs to allow the plate to heat up to steady state. To ensure there are no significant differences in the emissivity of the new plate compared to the calibration plate, the mean digital levels are compared to the mean steady state levels of the calibration plate to ensure the percent difference is less than 5%.

The dust delivery begins once steady conditions are then achieved. Simultaneous to starting the conveyor belt, the high-speed PSV video starts recording at a frame rate of 60 frames per second. An IR video is simultaneously started and set to record a still image every 5 seconds. The 1.4 grams of ARD are injected evenly over a 400 second period. When the dust has been delivered the belt is stopped and both videos stop recording. The facility is slowly cooled down and the target with the deposit structure is carefully removed and photographed. In this process, the deposit structure remains completely intact. The plate and deposit are weighed, and the clean plate mass is subtracted to yield a deposit mass. The facility is finally blown out and the residual dust stuck in the tubing is captured in a layered filter. The difference between the clean and dusted filter mass yields the mass that was not delivered to the target, which equaled approximately 0.4 grams ($\pm$ 0.00042 g). This amount is subtracted from the 1.4 grams injected to yield the dust delivered to the target, which equaled approximately 1 gram ($\pm$ 0.00052 g).

Chapter 18. Experimental Results

A top-down image of the final deposit cone is shown in Figure 18.1. The symmetry of the cone is notable and is a feature that allows for easier approximation of the 3-D volume of the cone from the 2-D PSV images. The bottom image in Figure 18.1 shows a corresponding contour of the deposit height normalized by the tube diameter (h/d). This data was extracted using a Keyence XR-40 3-D scanner. A 2-D trapezoidal integration of the deposit height data yields a deposit volume of approximately 266 mm^3. The measured mass of the deposit is 0.1872 g ($\pm$ .00042 g), and thus the effective deposit density is 703 kg/m^3 When divided by the nominal density of Arizona Road Dust (2727 kg/m^3), the result is a PF of 0.26. This value is required in the mesh morphing process as described previously and shown by Equations 13.3 and 13.4.

Figure 18.2 shows a PSV image of the clean plate at the very beginning of the test on the left, and on the right shows the fully-grown deposit shadow. Each image is first binarized using a specified threshold to distinguish the deposit shadow from lighter surrounding regions. In each image, the exit of the pipe is visible, and it is noticeable that a thermocouple was welded to the top of the outer surface of the tube to obtain a thermal boundary condition for the CFD. To obtain spatial measurements of the deposit, a calibration of the clean image is performed by measuring the number of pixels across the outer tube diameter in the image. The outer diameter measures 7.87 mm, which when divided by the number of pixels gives the distance-to-pixel conversion used to measure the deposit. To calculate the deposit height, the horizontal distance in pixels from the

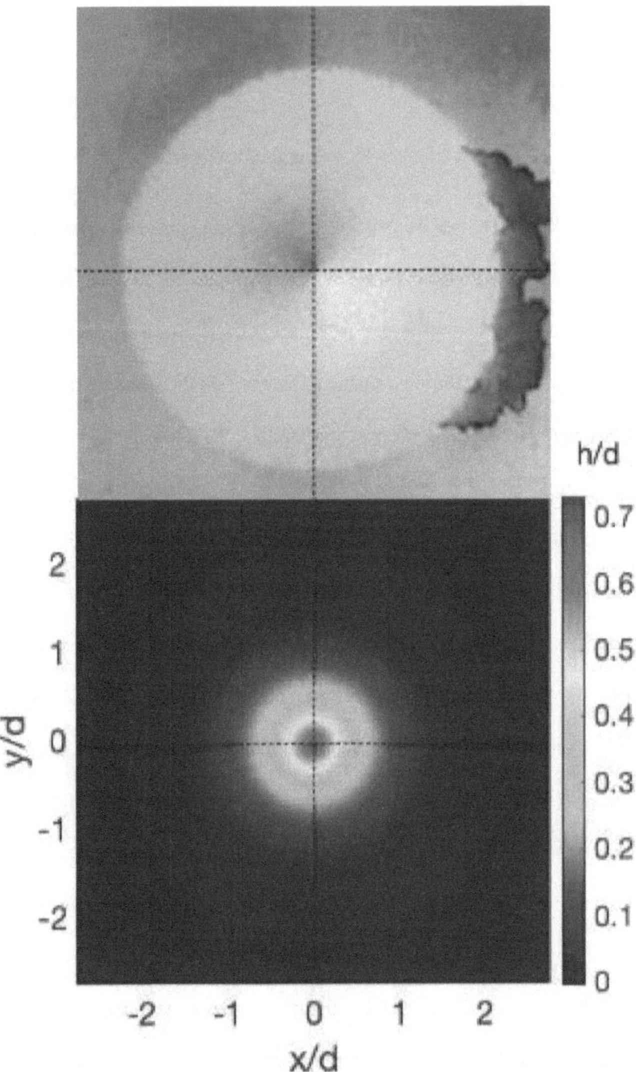

Figure 18.1: Top-down photo of final deposit cone structure (top) with normalized deposit height contour from 3-D scanner (bottom).

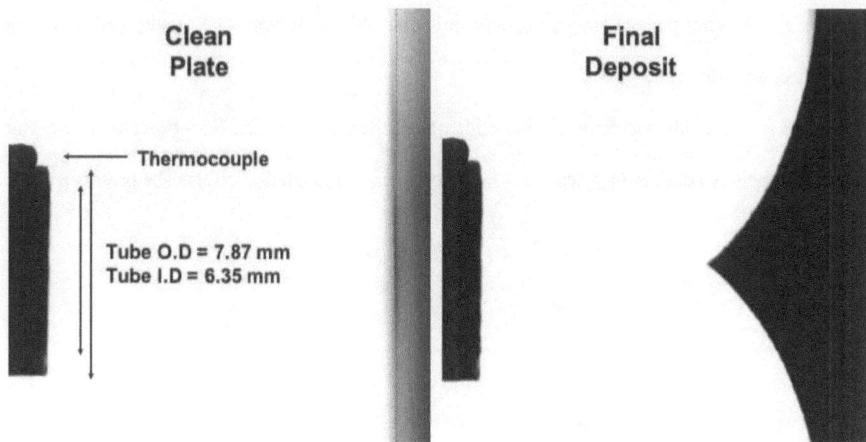

Figure 18.2: PSV images of the clean plate (left) and final deposit structure (right). Dotted red line indicates the clean plate surface used for measuring the deposit height.

clean plate surface to the outline of the deposit shadow is measured. Converting this to physical distance at the cone centerline yields the maximum deposit height. To approximate the volume, the deposit height is integrated both radially and circumferentially assuming that the cone is axisymmetric. The entire deposit cone is not within the frame of each image, and thus to make comparisons to the modeling results the volume integration is truncated to consider only the deposit within 2 tube diameters of the deposit centerline, which is entirely contained in each frame.

Figure 18.3 shows the evolution of the deposit within the truncated region. Each line represents 20 seconds of experimental time. The black dotted lines represent the region within the tube diameter. The deposit growth at the beginning of the test is clearly more gradual and fairly linear

121

with time and builds a base layer of deposit. Approximately halfway through the experiment when there is a significant plateau shape, the deposit begins to grow much more rapidly and in a non-linear fashion as the cone peak develops.

There is no heating source on the backside of the plate, and thus the flow heats the plate. The growth of the low conductivity deposit creates an increasing insulation effect. The presence of

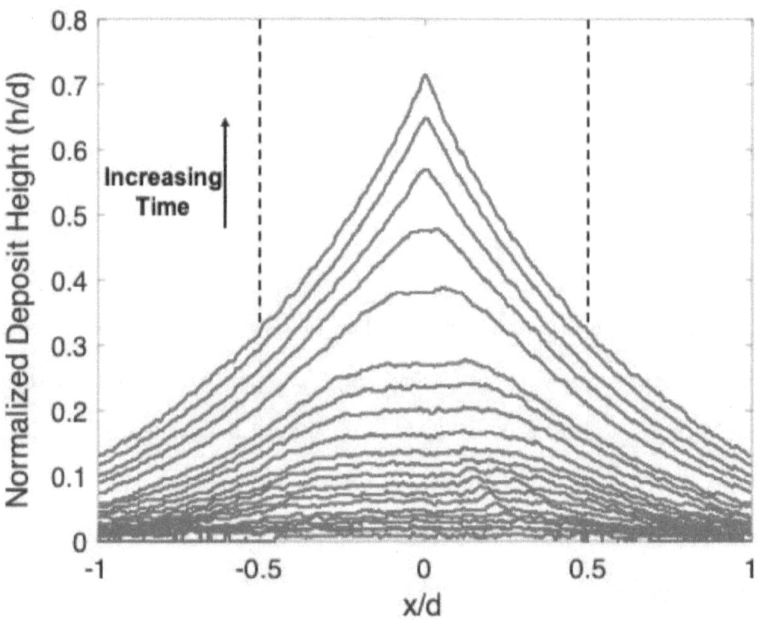

Figure 18.3: Deposit evolution from PSV images. Each line represents 20 seconds of experimental time. Black dotted lines represent the region inside the tube diameter.

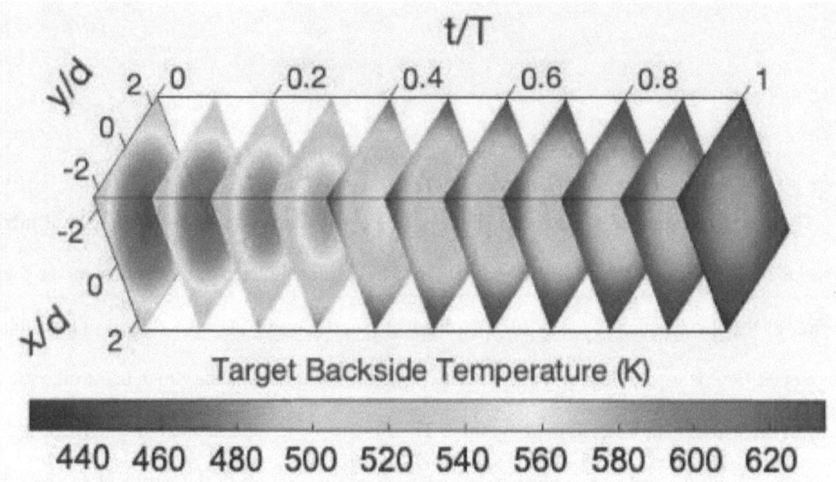

Figure 18.4: Thermal progression on the backside of the target as measured by IR versus non-dimensional time (t/T) at 40 second intervals. The plate area is cropped for increased clarity.

deposit then results in a drop in the temperature of the plate. The processed IR images are plotted as a function of the non-dimensional test time (t/T), where T equals 400 seconds and each increment is 40 seconds. To enhance the changes in the thermal profile as a function of time, the plate area is cropped to a square with side lengths equal to 4 tube diameters. The significant drop in the plate temperature as the deposit grows is prominent, and the maximum temperature at the centerline drops by approximately 25% over the duration of the test. While this thermal impact may be exaggerated compared to what occurs in an engine environment, this is ideal for evaluating the predictive ability of a conjugate mesh morphing framework as it reduces the influence of uncertainties in the experiment and the modeling.

Chapter 19. Conjugate Computational Setup

The conjugate domain is pictured in Figure 19.1 and a planar view of the tetrahedral mesh down the centerline of the tube is also shown. The grey region in the top left image shows the fluid region of the tube, which extends 20 tube diameters upstream of the tube exit to ensure fully developed flow is established at the tube exit. A mass flow inlet boundary matched to the mass flow of the experiment is set at the tube inlet. The flow rate and total temperature settings at the inlet, along with the rest of the relevant boundary conditions on all of the pertinent regions are detailed in Table 19.1. The inlet total temperature was not measured in the experiment, and thus is iterated in the computation until the centerline temperature at the tube exit matched the experimentally measured value of 894 K. The tube walls are not modeled as a solid region participating in conjugate heat transfer because the deposition and thermal impact on the tubes is not of interest in this study. The thermal boundary condition of the inner tube wall is set as a constant heat flux condition out of the pipe. This was approximated by first calculating the heat transfer coefficient using a pipe flow Nusselt number correlation by Gnielinski et al. [53]. The resulting Nusselt number is 11.94 and the corresponding heat transfer coefficient is 113 w/m^2-K. The inner and outer temperatures in the tube were measured experimentally at the tube exit and used to approximate the driving temperature difference for the heat flux calculation. The outlet is set to an ambient pressure outlet and is shown in red. The outlet extends sufficiently far from the edges of the plate in each direction to ensure the flow is diffused out well before reaching the outlet.

The solid target region is modeled, and the solid mesh is immersed in the fluid region as shown in the planar view of the mesh. The connectivity between the fluid and solid regions is established through a conformal interface. The mesh in the tube and in the region extending from the tube exit to the target has increased refinement to resolve the sharper flow and thermal gradients in this region. Each wall boundary in the fluid region consists of 5 prism layers to resolve the near wall gradients. The average non-dimensional wall distance (y^+) in the tube is 0.93 while the max y^+ of 1.96 occurs in the accelerating flow near the impingement location. The total mesh consists of 2,535,794 fluid cells and 399,028 solid cells.

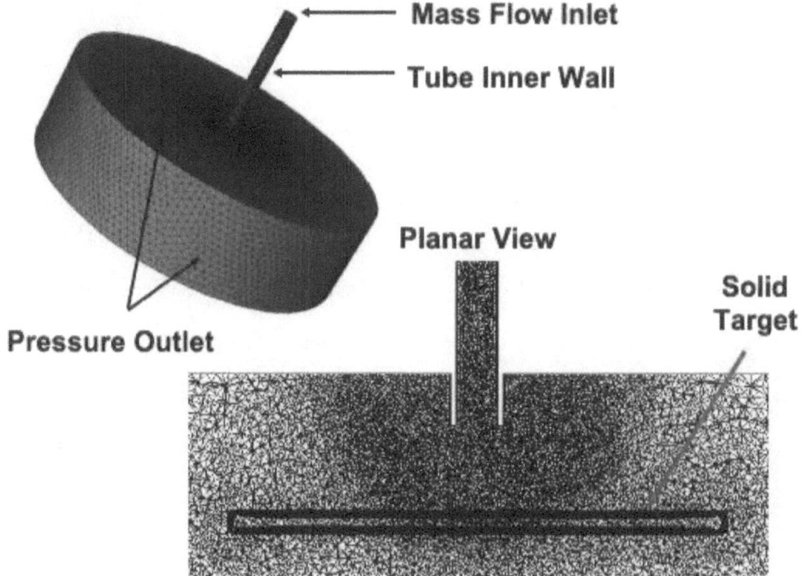

Figure 19.1: Computational geometry showing tube and outlet sections (top left). Planar cut-section of mesh at the pipe centerline zoomed in on the tube exit and solid target plate region.

125

Table 19.1: Boundary condition types and settings used in the conjugate CFD study.

Boundary Type	Boundary Settings
Mass Flow Inlet	$\dot{m} = 0.0007084\,\dfrac{kg}{s}$ $T = 914\ K$ $Turbulent\ intensity = 5\%$ $Turbulent\ Length\ Scale = 0.00023\ m$
Pressure Outlet	$P = 1\ atm$ $T = 298\ K$
Inner Tube Wall	No-slip $\dot{q} = -28,000\,\dfrac{W}{m^2 K}$
Outer Tube Wall	No-slip $T = 670\ K$

The steady flow-field was solved in ANSYS Fluent using a segregated pressure-velocity solver in addition to a segregated energy solver. The energy equation is of course solved for both the fluid and the solid domain in a conjugate simulation, while the pressure and velocity fields are solved only in the fluid domain. A RANS k-ϵ turbulence model with enhanced wall treatment was chosen to model the mean turbulence. Several models were investigated, and it was found that this model did the best job of capturing the convective mixing of the free shear layers that develop in the flow exiting the tube. This mixing is key in predicting the lateral heat transfer across the free shear layers as well as the eventual heat transfer on the target plate. The enhanced wall treatment uses a blending function that allows for the accurate prediction of the flow throughout different turbulence regimes in the near-wall region. This allows for more accurate resolution in regions where the y^+ exceeds 1 and the first wall cell lies in the buffer or transitional region of the boundary layer. The effect of gravity was included as there is a noticeable buoyancy effect that causes an asymmetric thermal profile in the experiment. A second-order accurate formulation is used for each equation solved by the CFD.

The fluid domain is given the properties of air under an ideal-gas assumption. The specific heat (c_{p_f}) and thermal conductivity (k_f) of the air are a function of temperature in Kelvin as given by Equations 19.1 and 19.2 respectively. The dynamic viscosity is calculated using Sutherland's Law. In a steady simulation, the solid region only requires the thermal conductivity to calculate the conduction through the solid. A temperature dependent conductivity is assigned as given by Equation 19.3, which is a fit to conductivity data of Hastelloy X from the open literature [54].

$$c_{p_f} = -5.108E\text{-}5T_f{}^2 + 0.27T_f + 916.1 \tag{19.1}$$

$$k_f = -1.1166E\text{-}8T_f{}^2 + 7.425E\text{-}5T_f + 0.00543 \tag{19.2}$$

$$k_s = 0.0193T_s + 4.075 \tag{19.3}$$

The steady state velocity field on the clean target plate is shown on a plane cut through the tube centerline in Figure 19.2. The figure clearly illustrates the ejection of the pipe flow into the free-stream, the developing free-shear layers, and eventual impingement and turning of the fluid when it encounters the front-side of the target. The maximum velocity occurs at the pipe centerline as expected and equals approximately 70 m/s. The average velocity in the pipe measured in the CFD is equal to 56.5 m/s as expected given the mass flow rate set at the inlet. It is evident that the jet is sufficiently diffused and has a negligible velocity at the outlet of the domain, ensuring the static pressure at the outlet is indeed equal to the atmospheric pressure.

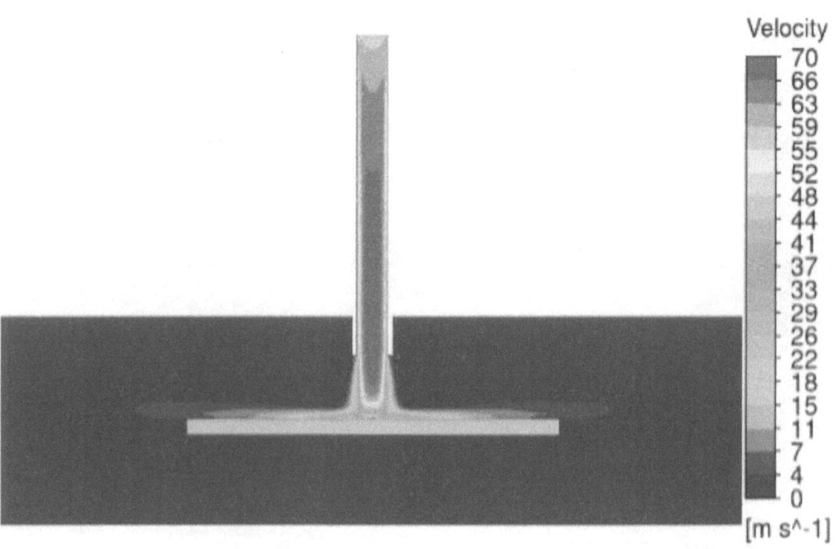

Figure 19.2: Steady state fluid velocity field on planar slice through the tube centerline.

The steady state temperature field on the same plane is shown in Figure 19.3. Although the resolution is not fine enough to see variations in temperature across the tube, the bulk temperature drops from the inlet to the outlet due to the heat flux through the tube wall. The centerline exit temperature is 894 K which matches the experimental measurement. In the exhausting free-shear jet, the conduction from the hot jet to the cool atmospheric air can be seen. The conduction through the Hastelloy X plate is prominent, and a thermal gradient is created from the center of the plate to the edges. This is due to both the lateral conduction through the solid as well as the loss of heat in the near-wall fluid as it turns and moves toward plate outer edges. This highlights the coupling of the fluid and solid thermal profiles that can only be full resolved via a conjugate simulation.

Figure 19.3: Steady state fluid temperature field on planar slice through the tube centerline.

Figure 19.4 compares the temperature contours on the backside of the target of the experiment before dust was injected to the steady state CFD solution. The experimental contour had to be cropped just slightly because the jet centerline was not exactly centered in the experiment. The image was cropped slightly to move it toward the center. Also, of note is at the bottom of the experimental contour, the metal bar that the target is clamped to is visible and thus obscures some of the image when compared to the CFD. Ultimately the deposit builds near the center of the plate, so any asymmetry that this might cause that isn't captured in the CFD is negligible on the results. The CFD contour shows reasonable agreement with the experimental result as measured by the IR. The region within approximately 5 diameters of the tube centerline is well predicted by the CFD. The regions further away from the jet centerline overpredict the temperature, which is

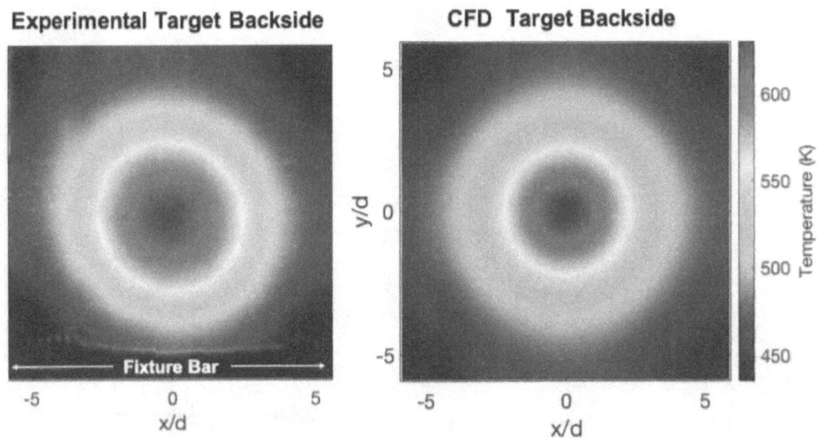

Figure 19.4: Steady state solid temperature field on the backside of the target plate for the experiment and CFD.

believed to be due to limitations in the ability of RANS turbulence models to accurately predict the wall heat flux in those regions. It was found that all of the RANS models available in ANSYS Fluent did a poor job of predicting this outer region of the plate. As stated previously, the PSV images have a limited field of view and thus the images are cropped to include only the region within 2 tube diameters of centerline before calculating the volume or maximum deposit height. The CFD conjugate morphing effort will also then focus on the dust that builds within this region as well as the change in temperature on the backside of the plate in this same area. The poor prediction occurs well outside this region and thus should have only a minor effect on the analysis.

Chapter 20. Conjugate Mesh Morphing Process

The routine for modeling deposition through mesh morphing in a conjugate simulation is fundamentally the same as was described in the flowchart in Figure 13.2 for morphing of a fluid domain. The first additional requirement is that the wall cells in the solid region must be displaced along the same vector as the adjacent fluid cells so that the vacated region of the fluid domain is filled in exactly by the solid. This is what occurs in a physical sense. To accomplish this, the conjugate mesh must use conformal interfaces between the fluid and solid regions. That ensures that for every fluid node in every boundary face, there is a solid node in the same exact location. In Fluent, the solid node corresponding to each fluid wall node is referred to as a shadow node. The number of faces on the fluid and solid wall interface are exactly matched and the meshes in each domain are an identical match. This makes morphing the solid straightforward, as all that is required is to identify the corresponding shadow node for each fluid boundary node and direct the dynamic meshing algorithm to move both nodes to the new coordinate as prescribed by the UDF.

The left image in Figure 20.1 shows what the mesh would look like if the fluid wall in the impingement region were morphed without also displacing the shadow nodes in the solid mesh. The fluid mesh is clearly deformed which would influence the solution in the fluid domain. Although there would be a gap between the fluid and solid walls, the nature of the interface would still allow the temperature field to be communicated and solved. The growth of the deposit in the solid region, however, increases the thermal resistance and affects the conduction. Failing to also deform the solid mesh would then miss the conductive length scales through the solid. The right image in Figure 20.1 shows the interface when the solid shadow nodes have been morphed. It is

clear that the solid domain has stretched to model the solid particulate that builds on the original

flat target, and the interface between the fluid and the solid maintains its conformality.

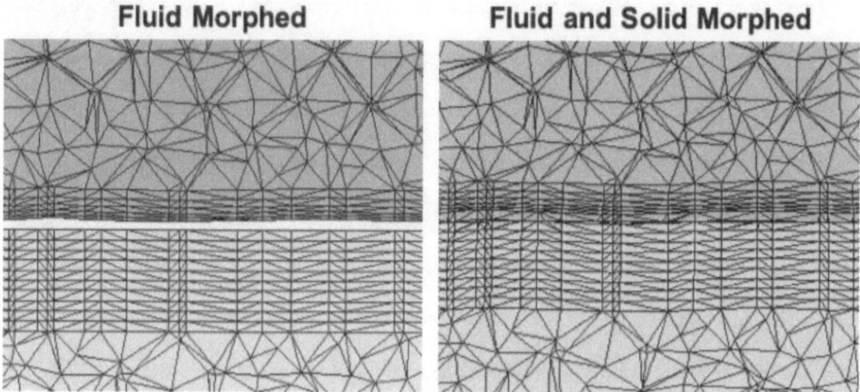

Figure 20.1: Morphing of fluid (red) target boundary only, resulting in a gap at the interface (left). Morphing of both the fluid (red) and corresponding solid shadow nodes simulating the deposit in the solid region (cyan) and maintaining the conformal interface (right).

To accurately model the conduction through the deposit and the metal plate, the cells that are

in the region of the solid that represents the deposit structure must be identified and differentiated

from the cells in the original plate domain. This is done very easily with this geometry since the

location of the frontside of the clean target is known. By looping over all of the cells in the solid

domain and checking the positions of each of the four nodes per cell, it is easy to check which cells

133

have a node that sits above the clean plate surface and thus are in the deposit region of the solid. It is common due to the unstructured nature of the mesh that cells have some nodes that lie above the plate and some that lie below in the plate region. Since the conductivity is assigned at the cell center and not at the nodes, it is necessary to make a determination on whether these cells will have the conductivity of the metal or the lower conductivity of the deposit.

One option is to assign a number weighted effective thermal conductivity to each cell, where the weighting of the metal conductivity contribution to the effective conductivity is based on the ratio of the number of nodes below the clean plate target surface to the total number of nodes in the cell. The weighting of the deposit contribution of course would be based on the fraction of the nodes above the target surface. For the modeling in this experiment, 10 thin prism layers were included in the wall regions of the solid mesh as can be seen in Figure 20.1. While prism layers provide refinement that is unnecessary in the solid region since it is simply a conduction problem, they add convenience in determining which prism cells are above and below the target. The thin nature of the prism cells makes it less likely that the nodes in those cells will straddle the interface boundary. In the following morphing results, if a cell has at least one node that lies above the original target surface, it is assigned the conductivity of the deposit, otherwise, it calculates the cell conductivity from the temperature dependent fit for Hastelloy X that was given in Equation 19.3. The resulting conductivity of the cells in the solid region after performing this check on the morphed solid is depicted in Figure 20.2. The top left image shows an isometric top-down view of the frontside of the target with a contour of the conductivity. The delineation between the metal and deposit is evident, as the deposit was assigned a much lower conductivity (0.1 W/m-K) than the metal. The bottom image shows a planar slice through the solid that cuts down the tube centerline. There is some apparent jaggedness where the dividing line occurs between the deposit and metal conductivity. This is a consequence of using an unstructured mesh, but the increased resolution helps minimize the impact.

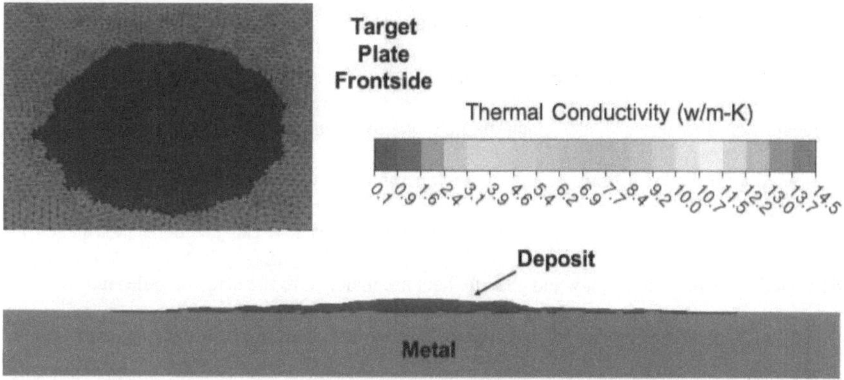

Figure 20.2: Contours of the solid thermal conductivity after morphing the mesh and assigning reduced conductivity to cells in the deposit region. Top-down isometric view of the target frontside shown in the top-left. Planar contour through the tube centerline on bottom.

Chapter 21. Conjugate Mesh Morphing and Deposition Settings

The mesh morphing procedure is identical to the one outlined for the effusion hole study in Chapter 13, only the solid domain is morphed in addition to the fluid domain. For this study, 12 diameters in the 0-5 µm range are injected into the domain. The values of those diameters were plotted previously in Figure 17.2 along with the cumulative mass distribution of the 0-5 ARD. For each size, approximately 14,900 particles are injected from an evenly spaced injector grid just inside the tube inlet. This number was found to yield statistically converged number capture efficiencies on the target plate. Each particle is given an initial velocity and temperature corresponding to the local flow velocity and temperature. The forces affecting the particle in the DPM model match that of the earlier mesh morphing effort and can be found in Chapter 13. Additionally, the particle density and specific heat are matched to the effusion hole study.

The mechanical properties for the impact physics to determine the sticking and rebound behavior were also matched to the previous study and can be found in Table 13.1. The only exception is that the surface free energy (γ) was lowered to a value of 0.1, as this provided the best match in terms of the overall capture efficiency and resulting deposit evolution as a function of grams delivered. The comparison of the computational and experimental evolution will be provided in the next Chapter. It is unlikely that γ is a function of the particle diameter, which is the only apparent difference between the 0-10 and 0-5 µm distributions of ARD. This is likely because the work of adhesion in the OSU Deposition Model is calculated by multiplying γ by the contact area of the particle and the substrate. This is a common formulation used in tribology, and

γ is treated as an energy flux, or independent of the particle diameter. That said, it is the easiest constant to tune in the OSU deposition model to control the sticking probability. The need to reduce γ so significantly (from 0.8 to 0.1) does however lead to some concerns about the generality of the model that will need to be vetted in future research.

The key question to be answered is what physics or properties are fundamentally different between the effusion hole mesh morphing study and this impinging jet study such that the same mechanical properties cannot be applied for both and give the desired accuracy. It is likely that in this instance, this reduction in γ from 0.8 to 0.1 is accounting for a change in a different mechanical property that is unclear at this time. One candidate is that particles in the impinging jet impact the deposit cone at angles that make them less prone to embedding themselves in the porous voids of the surface. This effect is not explicitly considered by the model and would likely have to be handled empirically. The model also makes the assumption that every impacting particle is a cylinder impacting head on and experiences the full range of its maximum elastic-plastic energy exchange while the contact area is maximized. This greatly reduces complexities and computational requirements of the model but may have some limitations. For example, if the particles have differences in their sphericity or shape as a function of diameter as suggested by Figure 10.5, the contact area at impact may be different and change the adhesion forces. Figure 10.5 suggests that the smallest particles in a distribution of 0-10 µm ARD are more spherical than larger diameters. Since the model assumes that the contact area of the particles scale strictly with the particle diameter squared by assuming cylindrical particles, it may be that neglecting non-sphericity underestimates the scaling of the contact area. The contact area calculation in the model could be simply multiplied by the inverse of the sphericity as defined in Equation 10.5 to provide this functionality, however, the magnitude of that adjustment will still be 2-4 times smaller than the factor of 8 change in γ that is needed to accurately model the deposition in this study. Finally, consideration should be given to the treatment of ARD and other dusts as homogenous mixtures with effective properties that are size independent. While the volume fraction of the different chemical constituents of ARD are provided by PTI, there is no readily available data that clarifies if these are evenly distributed by size or if smaller particles tend to be composed of one particular constituent. If different subranges of diameters tend to have very different compositions, the assumption of composite mechanical properties that are size independent could cause significant errors in the modeling. While accurately modeling the growth is a priority, the means of doing so and the accuracy of the physical properties fed to the model is not of particular concern for this study.

10 mesh morphing iterations are performed, each accounting for 20 seconds of experimental time. Only the first half of the experiment is modeled because each mesh morphing iteration is more computationally demanding than the effusion hole morphing study due to the larger mesh. Since 1 gram of dust was evenly delivered over the duration of the experiment, each iteration in the mesh morphing procedure accounts for 0.05 g injected. The PF is set to 0.26 to match what was measured from the final experimental deposit. Although dust does build on the tube walls in the experiment, only the frontside of the target was morphed as this is what dictates the thermal progression on the backside of the plate.

After the fluid and solid regions are morphed, the conductivity of the deposit cells must be assigned as discussed previously. The conductivity of the porous deposit structure cannot be readily obtained from any measurements in the experiment, so to start with a value was estimated using information from the open literature. The value of the effective conductivity (k_{eff}) for a quartz structure with porosity of 0.7, or PF = 0.3, was measured by Kersten et al. in 1949 [55] and was approximately 0.1 W/m-K. Typical ARD distributions have a chemical composition that is more than 70% quartz by mass, so for the first modeling attempt, this value is used for cells determined to be in the dust region. Once the conductivities have been reassigned to the newly morphed geometry, 1000 iterations of the solver are run to recalculate the solution.

Chapter 22. Conjugate Mesh Morphing Results

The first conjugate mesh morphing simulation uses an effective deposit conductivity value of 0.1 W/m-K in order to determine how accurately the thermal prediction as a function of total grams delivered matches the experimental progression. To emphasize the importance of delineating between the metal and deposit cells and assigning realistic conductivity values to both, the same mesh morphing procedure is run without reducing the conductivity of the cells in the deposit region. In this case, the entire solid mesh, even the regions above the original surface that represent the deposit structure, are assigned the conductivity of Hastelloy X. In the plots and discussion to follow, this case will be referred to as "k_{metal}."

Figure 22.1 plots the normalized peak deposit height (h/d) versus t/T for both computational mesh morphing cases against the experiment. It is seen that both of the morphing simulations show good agreement with the experimental trend. The growth is roughly linear in the first half of the experiment and the mesh morphing predicts the same trend. It is also seen that the k_{metal} peak deposit heights are only slightly lower than the $k_{eff} = 0.1$ case. Both cases use the same mechanical properties for the impact, none of which are a function of the plate or deposit temperature at impact, so it is expected that the sticking predictions would be approximately equal. The slight difference between the two cases is most likely due to inherent randomness in the particle trajectories caused by the Discrete Random Walk model.

Figure 22.2 plots the deposit volume within 2 tube diameters of the centerline for the same three cases. As noted previously, the experimental volume is obtained from the PSV cone profiles using a trapezoidal integration of the deposit height and assuming the cone is axisymmetric. The volume of the cones built through mesh morphing is similarly calculated by performing an area integral of the deposit height of each node on the frontside of the solid target mesh. The volume trends for each of the computational morphing cases overlap and again are linear. The slopes are slightly overpredicted, but the trends converge upon the experimental volume at $t/T = 0.5$. Further iterations were not performed to determine whether the non-linearity seen in the second half of the experiment would be captured by the CFD, or if the linear behavior would continue. This is a worthwhile investigation for future research efforts but would either require the morphed geometry to be exported and completely re-meshed, or improvements to be made to the local re-meshing settings in the Fluent dynamic mesh to avoid the creation of highly skewed cells in the latter iterations.

Figure 22.3 compares the experimental evolution of the deposit during the first half of the experiment to the corresponding evolution modeled using mesh morphing. The height of the deposit within 1 tube diameter is very well predicted by the model at each 20 second increment plotted. The maximum deposit height which was plotted in Figure 22.1 occurs near the centerline and as such also shows close agreement to the experiment. Between 1 and 2 diameters however, the model overpredicts the deposition, which results in the overprediction of volume that is shown in Figure 22.2. One explanation that could explain why particles stick at a higher rate in this region in the modeling is that the model's sensitivity to the impact angle of the particle could be inadequate. Another could be that the isotropic assumption of the turbulent kinetic energy inherent to the k-ϵ turbulence model causes particles in the tube to distribute disproportionately closer to the tube walls. This is a known consequence of using the DRW model with isotropic turbulence models as described by Forsyth et al. [43].

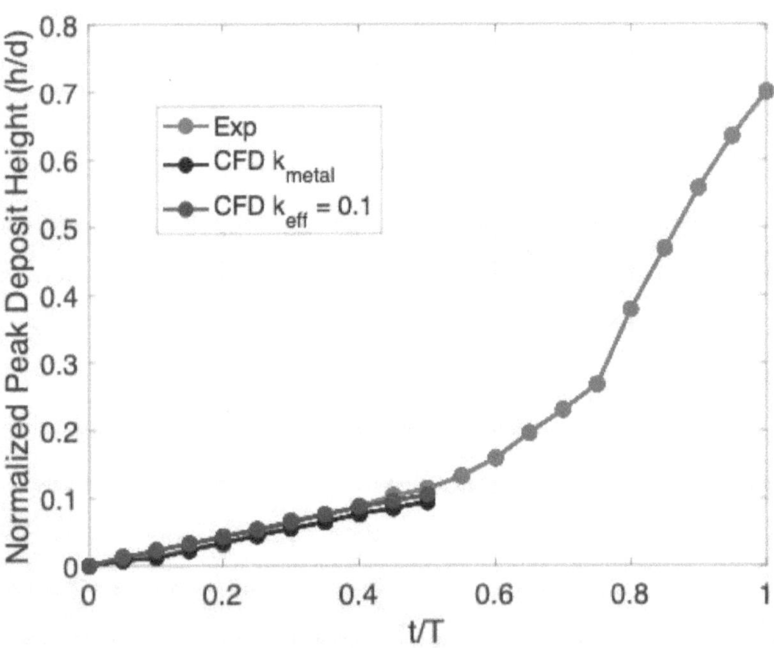

Figure 22.1: Normalized peak deposit height (h/d) versus t/T for the experimental cone growth (red), mesh morphing with metal conductivity throughout the solid (black), and mesh morphing computation with $k_{eff} = 0.1$ in deposit cells (blue).

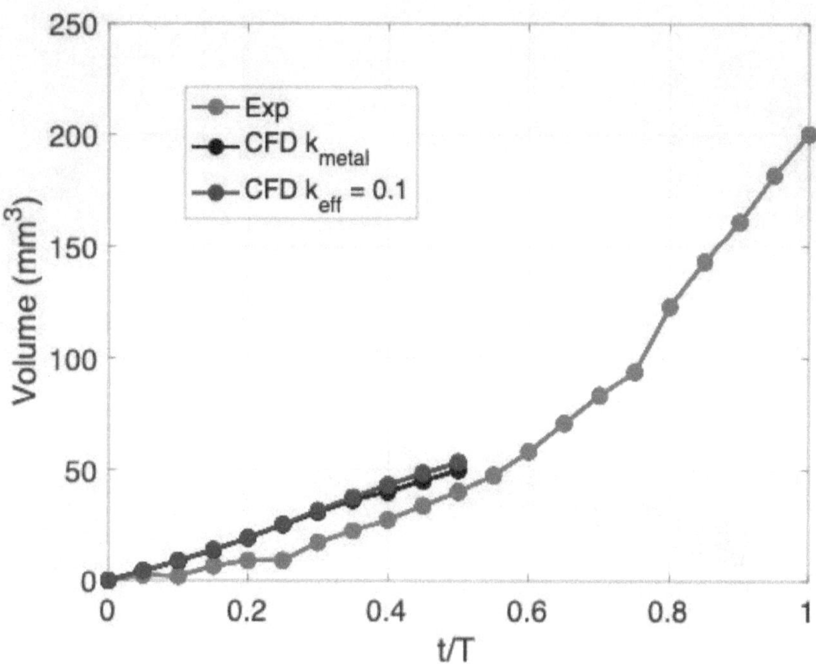

Figure 22.2: Deposit volume within 2 tube diameters of centerline versus t/T for the experimental cone growth (red), mesh morphing with metal conductivity throughout the solid (black), and mesh morphing computation with $k_{eff} = 0.1$ in deposit cells (blue).

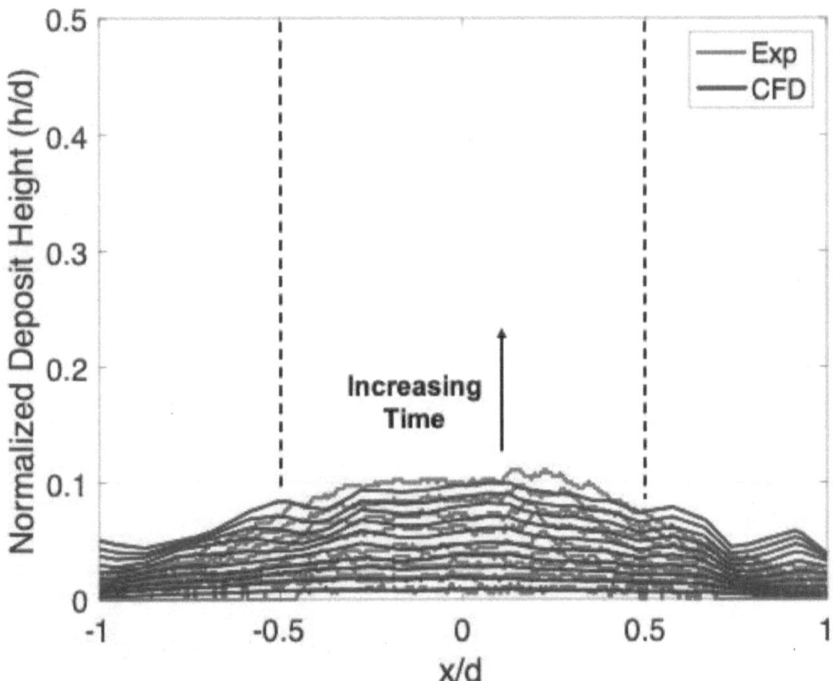

Figure 22.3: Deposit evolution comparison between CFD ($k_{eff} = 0.1$) and experiment up to $t/T = 0.5$. Each line represents 20 seconds of experimental time. Black dotted lines represent the region inside the tube diameter.

The temperature contour on the backside of the target plate as a function of t/T for the $k_{eff} = 0.1$ case is plotted in Figure 22.4. The result shows that the buildup of the deposit in the CFD, combined with the assignment of a decreased conductivity to the deposit cells, results in a steady

144

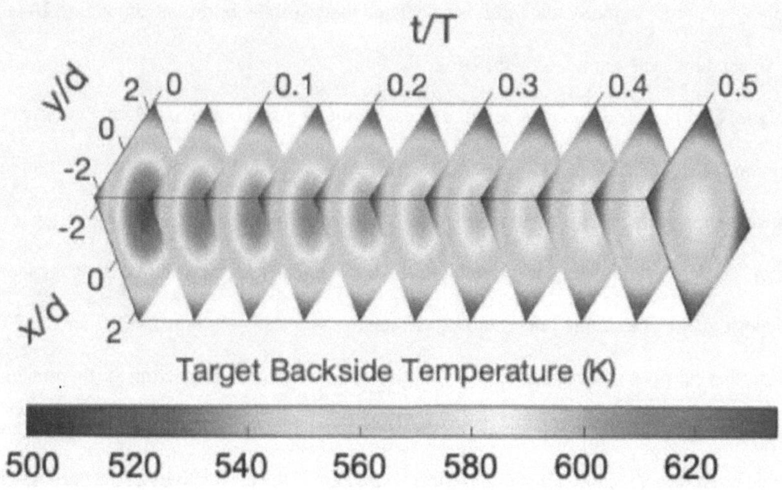

Figure 22.4: Thermal progression on the backside of the target versus t/T at 20 second intervals as predicted by the conjugate mesh morphing simulation with k_{eff} = 0.1 W/m-K. The plate area is cropped for increased clarity.

decrease in the backside target temperature. This alone shows the ability of the conjugate mesh morphing framework to be used for thermal predictions for the solid surfaces that deposits accumulate on. It does not, however, make any direct comparison to the experimental thermal progression in order to determine if the prediction accurately replicates the real physics.

To make this comparison, the region of the backside of the target within 2 tube diameters of the centerline was divided into two zones. The first zone, Z1, encapsulates the area of the target within 1 tube diameter of the centerline. The second zone, Z2, embodies the remaining area between 1 and 2 tube diameters. A simple diagram of these zones is provided in the bottom of Figure 22.5. In each zone, the mean temperature is recorded for each iteration of the mesh morphing solution. Similarly, the mean is calculated in each zone for the corresponding 20 second experimental intervals captured by the IR video.

Figure 22.5 plots the mean temperature in each zone as a function of t/T for both mesh morphing cases ($k_{eff} = 0.1$ and k_{metal}) versus the experimental data. For each case and at all times, the mean temperature in Z1 is higher than in Z2 as expected due to the nature of impinging jet heat transfer. The k_{metal} simulation has virtually no change in the mean temperature despite the fact that the deposit grows at a similar rate as the experiment, as was shown in Figures 22.1 and 22.2. This suggests that simply modeling the deposit growth is not sufficient for predicting the thermal impact because the changes in the fluid domain around the cone play a secondary role in the heat transfer. The first order effect is caused by the significant reduction in the conductivity of the porous deposit compared to the metal surface. It is imperative to model this phenomenon if the deposition model with the conjugate mesh morphing framework is to be utilized for predictive or design purposes.

By reducing the conductivity of the deposit to 0.1 W/m-K, it is seen that mean temperature drops monotonically in both zones. While the initial reduction in the mean temperature shows reasonable agreement with the experimental trends, the behavior for $t/T > 0.2$ diverges significantly from the experiment. The slope in this region is much shallower than the experiment, and the mean temperatures at $t/T = 0.5$ are approximately 50 K higher. Since the growth of the morphed deposit in time shows reasonable agreement to the experiment, it follows that there is some physics in this latter portion of the experiment that is responsible for the more dramatic drop in the backside temperature that is not being modeled appropriately. The most logical candidate worth investigating is the assignment of a constant value of the effective conductivity of the porous deposit.

The effective conductivity of the deposit structure is applied as a homogenous property with

no variation spatially in the deposit or in time as the deposit structure grows. While the

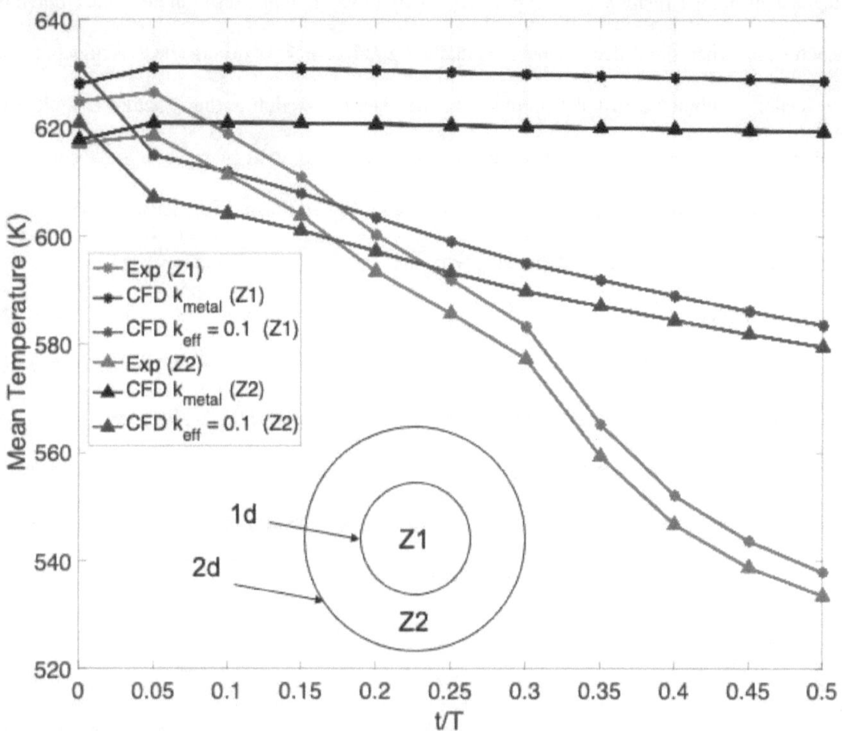

Figure 22.5: Mean temperature in zone 1 (Z1) and zone 2 (Z2) on the backside of the target plate versus t/T for the experiment (red), mesh morphing computation with metal conductivity throughout the solid (black), and mesh morphing computation with k_{eff} = 0.1 in deposit cells (blue). Zonal diagram provided (bottom).

147

conductivity of a pure solid is indeed an intensive property, in the case of a porous structure immersed in a fluid jet, the effective conductivity can change as the structure grows. This is because the effective conductivity is accounting for convective heat transfer due to the fluid flow through the porous structure as well as pure conduction through the solid particles and regions of stagnant air in the structure. The effective conductivity value should result in the same amount of heat transfer when simplified to a pure conduction problem as what occurs due to conduction and convection combined across the porous structure. The convection in the structure is limited to regions where the fluid jet penetrates and has a non-zero velocity. Regions closer to the surface of the deposit will of course experience a larger heat transfer coefficient as the local velocities in the porous regions will be higher. As the fluid penetrates the porous structure, however, it will diffuse and exit radially through the sides of the structure. When the deposit is sufficiently large, the regions near the base of the deposit will experience a very minimal amount of convection because the jet penetration depth is smaller than the thickness of the structure. In these areas, it reasons that local heat transfer simplifies to a pure conduction problem. If the penetration depth of the jet is constant as the deposit develops, it is logical that k_{eff} will be at a maximum when the deposit thickness is at a minimum, and a significant amount of convective heat transfer occurs throughout the structure. As the structure grows, k_{eff} will decrease and asymptote to a constant value when the base layer experiences zero convective heat transfer.

To illustrate this concept, another CFD simulation is run. A new geometry and mesh are created that model a canonical deposit cone with radius and height equal to that of the final deposit cone in the experiment. The cone sits on the same target plate and is centered directly under the impinging fluid. A planar view of the new mesh is shown in Figure 22.6. This simulation is not a

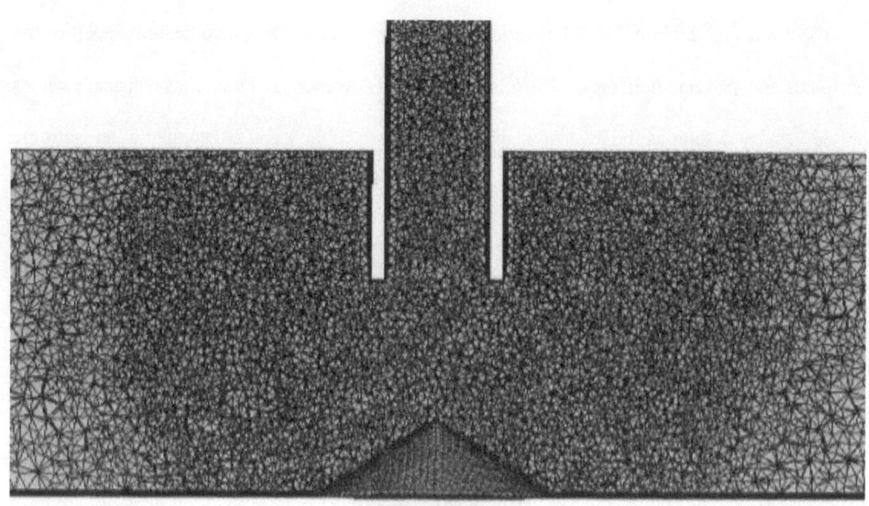

Figure 22.6: Modified mesh with deposit cone (red) modeled as a porous region using ANSYS Fluent porous model.

conjugate one, and only requires modeling of the fluid domain. As such, the thickness of the solid target and the stagnant fluid on the backside of the target are unnecessary to model. The same boundary conditions are applied as in the mesh morphing study. The objective is to visualize the penetration of the jet into the porous cone by using the ANSYS Fluent porous model. This requires that the cone is meshed and prescribed as a fluid region and set as a porous zone. The porous model uses empirically determined flow resistance and applies a sink term in the governing momentum equations to cells within the porous zone. The momentum sink contributes to the pressure gradient in the cell that decelerates the fluid.

The porous model in ANSYS Fluent includes a conical option, in which the user specifies the cone axis and the cone-half angle, which is 31.5° for this geometry. The viscous resistance values are defined by default as the inverse of the absolute permeability and are applied in an isotropic fashion. The only other variable required for the momentum sink calculation is the porosity, which is simply equal to (1-PF), or 0.74 for this simulation. These settings are all used in a relative velocity resistance formulation to calculate the pressure drop through the medium. The model solves the energy equation in the porous region by invoking an effective thermal conductivity that is a volume weighted average of the fluid and solid conductivities. While this seems like a similar application to what is used in the mesh morphing conjugate solution for the solid heat transfer, the difference is that the cone in this case is modeled as a porous fluid. As such, the advection term in the energy equation is non-zero and allows for the bulk transport of heat throughout the cone structure. That in combination with the effective fluid conduction allows the convection to be approximated in the deposit.

The velocity contour on the same centerline plane using the porous model is provided in Figure 22.7. As the jet impinges on the deposit cone, the majority of the fluid mass is decelerated and turned down the sides of the cone surface. A portion of the fluid mass near the cone peak however penetrates into the structure. This portion of the mass flow diffuses through the voids of the structure and exits through the sides of the deposit. The convective heat transfer will of course be highest near the peak of the cone, and virtually negligible near the base. Although this simulation uses a first order empirical model with a handful of limiting assumptions, the result illustrates the general manner in which the fluid penetrates into the structure. It is logical that if the height of the structure were decreased, the fluid convection would be more aggressive over a larger fraction of

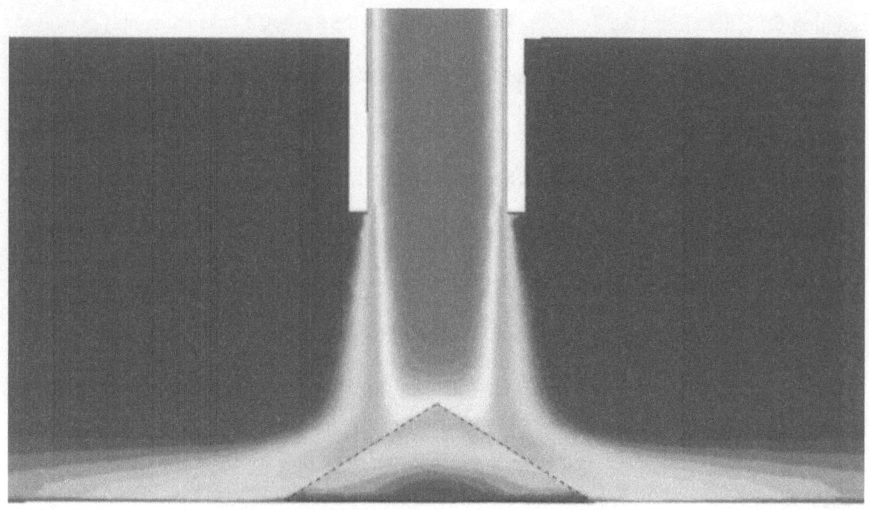
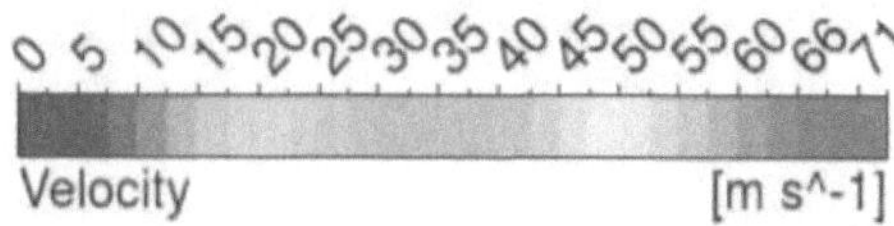

Figure 22.7: Velocity contour at centerline plane using the porous media model to approximate the flow through the deposit cone.

the cone area. To model this augmentation in convection in the mesh morphing routine where the heat transfer through the deposit is treated as pure solid conduction, the effective conductivity must be increased.

To apply this principle of an effective deposit conductivity that decreases as the structure grows, the results in Figure 22.5 for the $k_{eff} = 0.1$ case are first corrected to match the mean temperature behavior of the experiment. Realizing that the development of the deposit is nearly independent of the value of the solid conductivity and the corresponding solid temperatures, the solution for 5 iterations (1st, 2nd, 5th, 7th, and 10th) of the $k_{eff} = 0.1$ case are simply re-solved. New values of k_{eff} are iterated through until the mean temperatures in each zone match the experiment within 5% at the corresponding t/T. The new values of k_{eff} are then plotted as a function of the maximum deposit height for each iteration and a 2nd order polynomial is fit to the data as shown in Figure 22.8. K_{eff} asymptotes to a value of approximately 0.04 W/m-K in the generated curve fit. In Figure 22.8, the conductivity of stagnant air at two different temperatures are also plotted. The conductivity of air at centerline of the exhausting jet (894 K) is approximately 0.063 W/m-K according to Equation 19.2. The air inside the deposit structure is coolest at the base of the deposit, where the air temperature drops to 670 K. The conductivity in this region is approximately 0.05 W/m-K. Comparing the values from the curve fit to the conductivity of air at these two temperatures reveals two things. First, when the deposit thickness is maximized the heat transfer is dominated by pure conduction through the deposit, which is primarily air by volume. The minimum k_{eff} required to match the experimental mean temperatures is 0.04 W/m-K, which is very close to the stagnant conductivities of the air inside the deposit. This demonstrates that the required values make physical sense which gives confidence in the fidelity of the conjugate mesh morphing procedure implemented. This value of 0.04 W/m-K is however slightly lower than the conductivity of the stagnant air inside the deposit, which is an unphysical outcome. This could very well be due to the lack of radiation modeling. The deposit may dissipate a non-negligible amount of heat via radiation to the surroundings. Since radiation isn't modeled explicitly, k_{eff} may need to be reduced even more than what makes sense physically to account for the radiation and to match the experimental trends.

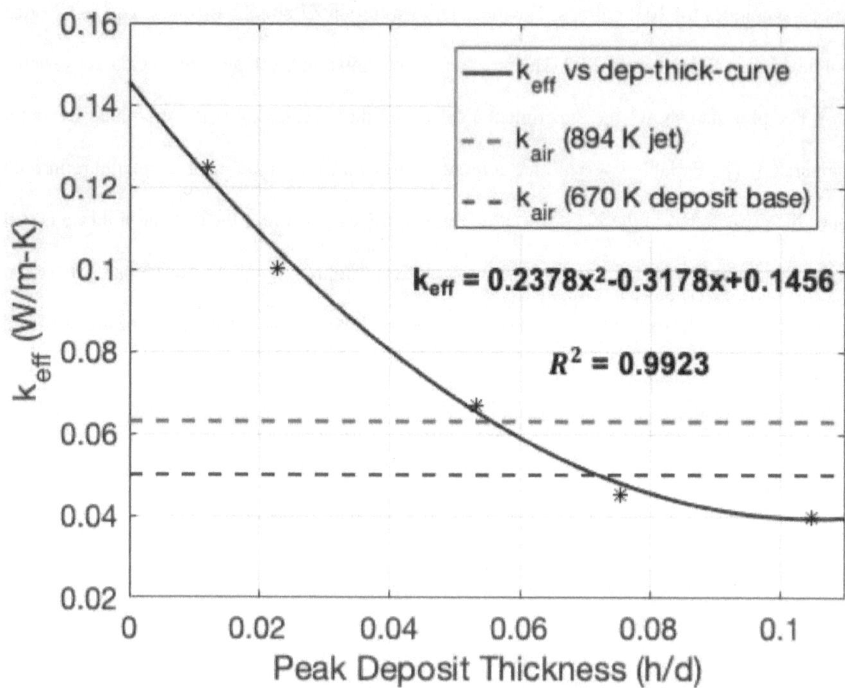

Figure 22.8: 2nd order polynomial curve fit of k_{eff} vs maximum deposit thickness after re-solving the flow for 5 iterations of the $k_{eff} = 0.1$ case with new values of k_{eff} to obtain a closer match to the experimental mean temperatures.

The mesh morphing simulation is re-run using the curve fit from Figure 22.8 to determine k_{eff} for the deposit cells. After morphing the domain, the peak deposit height on the target surface is found and plugged into the fit to yield k_{eff} for that iteration. The conductivities are then assigned to the dust and the metal respectively, and the solution is re-calculated, and the mesh morphing process continues for 10 iterations. The mean temperature in Z1 and Z2 are calculated and plotted as a function of t/T in Figure 22.9. The new trends are shown in black and are labeled as "variable k_{eff}." The plot also shows the experimental trend and the original $k_{eff} = 0.1$ morphing trend for comparison. The variable k_{eff} trend has a more significant drop in the mean temperature in both zones than was found in the $k_{eff} = 0.1$ case. The cooling as a result of the deposit build-up is still slightly underpredicted, but the result shows significant improvement. The discrepancy between the new trend and the experiment can most likely be attributed to the simplicity of the method used to assign k_{eff}.

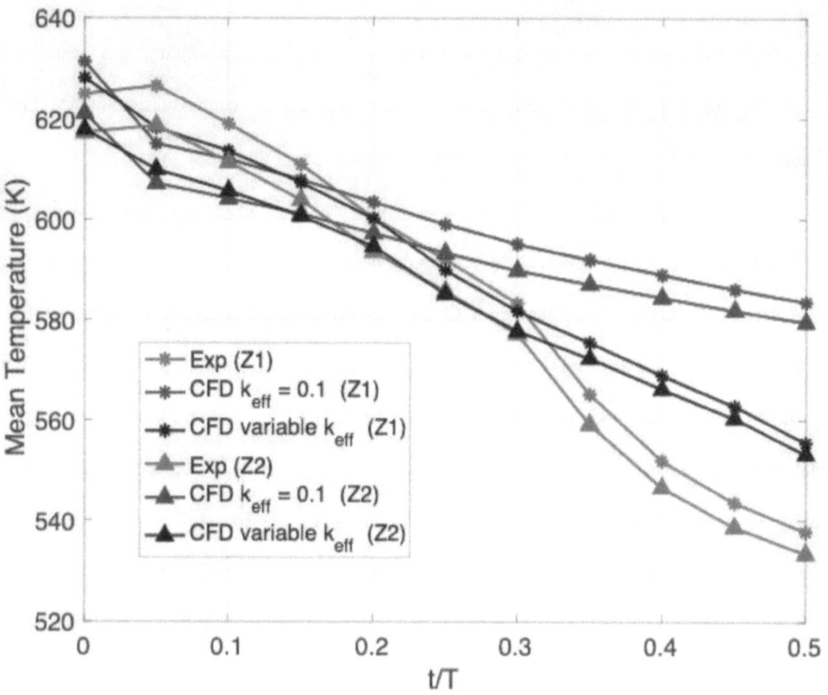

Figure 22.9: Mean temperature in zone 1 (Z1) and zone 2 (Z2) on the backside of the target plate versus t/T for the experiment (red), mesh morphing computation with $k_{eff} = 0.1$ in deposit cells (blue), and mesh morphing computation with variable k_{eff} (black).

155

Chapter 23. Conclusions Regarding Conjugate Mesh Morphing

A deposit cone was formed using an oversized impinging jet geometry on a Hastelloy X target to provide a case study where a significant thermal response occurred due to the insulative effect of the particulate. 1 gram of 0-5 ARD was delivered over a 400 second period and the deposit growth was visualized with a PSV technique over the duration of the test. Infrared imaging was used to capture video of the temperature field on the backside of the target. The PSV images were sampled every 20 seconds and binarized to identify the outline of the deposit. A distance-to-pixel calibration was applied to plot the deposit height of deposit profile at each interval. The growth was shown to be slower and linear during the first half of the experiment and a shallow plateau structure formed. In the second half of the experiment, the dust accumulated more rapidly and non-linearly as the cone peak filled out. The thermal images were post-processed at matching intervals and the progression was plotted. The target temperature showed a steady decline as the deposit thickness increased, reaching approximately a 25% reduction of the centerline temperature at the conclusion of the test.

The mesh morphing framework was then extended to allow for the deformation of the solid domain in a conjugate simulation. This was simply performed by creating conformal interfaces where the fluid and solid meshes coincide. This ensures that each fluid boundary node has a corresponding "shadow node" on the solid boundary in the exact same location. Each shadow node in the solid is morphed along the same vector as it's matching fluid boundary node, which guarantees the solid mesh exactly fills in the vacated space from the fluid mesh deformation. To accurately model the heat transfer in the morphed solid domain, the deposit and metal cells within the solid mesh are first identified based on their position relative to frontside target surface. They are then assigned either a thermal conductivity of metal or an effective thermal conductivity of the deposit, which is significantly lower than that of the metal.

This conjugate mesh morphing ability was tested on a conjugate domain that replicated the experimental impingement flow. The initial steady state solution before dust delivery was solved and the temperature field on the backside of the target was found to show close agreement with the experiment within 5 tube diameters of the centerline. Each iteration of dust delivery and mesh morphing simulated 20 seconds of the experimental time and thus 0.05 g delivered. 10 iterations were performed, simulating the first half of the experiment.

The first case tested used an estimated effective thermal conductivity value of 0.1 W/m-K based on a study of a quartz deposit with porosity of 0.7. The second case prescribed the conductivity of Hastelloy X to the entire solid region, even the cells in the deposit region. The volume with two tube diameters of the centerline and the normalized peak deposit height at each iteration of the two computational cases were plotted against the corresponding non-dimensional time in the experimental progression. Each of the simulations grew at approximately the same rate and showed reasonable agreement with experimental evolution (by adjusting γ). The mean temperatures were calculated in two concentric zones within two tube diameters of the centerline for both computational cases at each iteration and plotted against the corresponding experimental data. The case run with metal conductivity throughout the entire solid showed virtually no change in the mean temperature despite showing realistic deposit growth. This stressed the importance of assigning reduced conductivity to the deposit cells in making accurate thermal predictions.

The case using an estimated effective conductivity of 0.1 W/m-K resulted in a monotonic decrease in the mean temperature in each zone on the target backside, but the thermal impact was underpredicted compared to the experiment. Since the growth of the deposit is modeled fairly accurately, the most likely candidate that explains the underprediction of the thermal resistance for conductive heat transfer is the use of a constant effective conductivity for the deposit regardless of the deposit thickness. To demonstrate why the effective thermal conductivity is likely a function of the deposit thickness, a new computational mesh was created that modeled a fully developed cone directly under the impinging jet. This simulation did not model conjugate heat transfer and was intended to illustrate how the fluid jet penetrates the porous structure. The ANSYS Fluent porous zone model was used to model the flow through the deposit and the result showed that the velocities of the fluid in the structure are largest near the cone peak, and negligible near the cone base because the mass diffuses out of the sides of the cone before reaching the base. It was explained that these regions with higher velocities will have larger convective heat transfer coefficients, and regions with little to no fluid penetration reduce to a pure conduction problem. Logically, thin deposit structures will see a greater fraction of their internal surface area heated by convection and will have a correspondingly higher effective thermal conductivity than tall deposit structures.

This principle was applied to improve the conjugate mesh morphing prediction. An effective conductivity versus maximum deposit thickness curve fit was created by iterating through values of k_{eff} and re-solving the solutions for 5 iterations of the original mesh morphing case until the results showed closer agreement with the experiment. Using the 2nd order polynomial fit, k_{eff} was treated as a variable that decreases as the deposit grows. The resulting mean temperature trend showed much better agreement with the experiment than the constant k_{eff} case.

Although the result was improved, the application of the variable k_{eff} case was performed in a simplified manner to demonstrate the principle that the conjugate mesh morphing framework developed can provide accurate prediction capabilities if the correct properties are known and applied. In the current state, the user must have a more thorough understanding of the effective conductivity of the structure and how it changes as a function of not only the deposit shape, porosity, and chemical composition, but also the characteristics of the penetrating fluid.

Chapter 24. Future Work

Future efforts to improve the specific predictive capability for the thermal impact of deposition should involve targeted studies to better understand both the mechanical and heat transfer properties of the porous deposit structures. To move toward a more universal prediction capability, the deposition community should consider merging existing porous zone modeling into the existing framework. This would likely require the ability to separate the metal and deposit cells within the solid domain into two distinct zones after morphing the domain. The deposit zone would be specified as a porous zone, and the emphasis would be on specifying the inputs to the porous model to tune it for deposition applications.

One element of the current framework that was underutilized in this study but is critical for accurately predicting deposition is identification of deposit cells versus the original metal cells. This delineation was made in order to reassign more realistic thermal conductivity values as was necessary for capturing the experimental thermal trends. This functionality, however, is critically important for predicting the different sticking behaviors of particles impacting deposit and particles impacting bare metal. The modeling in this work did not make this distinction because there is a general lack of experimental data to inform the mechanical properties for these two different scenarios. The mechanical properties as tuned for the OSU Deposition Model have been empirically tuned to provide sticking rates and deposit growth that match deposition trends for common but specific flow-fields. To truly make strides toward universality, the mechanical properties for particles impacting existing deposits must be more rigorously studied and applied to the model. The conjugate morphing framework as constructed will readily allow for this added sophistication and fidelity.

Both the effusion and impingement mesh morphing experiments used a coolant flow that was hotter than the backside of the test plate. In the case of the effusion plate, this was necessary because of the challenges involved with providing an external heat flux to the backside of the effusion plate inside a pressure vessel. In the impingement scenario, the backside heat flux could have been applied with a torch, however this would have obstructed the view of the IR camera on the backside of the target and corrupted the digital levels measured by the camera. While the objectives of each morphing study were not affected by this deviation from the role of the coolant in real engine operation, the manner in which the effective conductivity of the deposit is modeled in the conjugate framework may be affected.

In the impingement study conducted it was demonstrated the effective conductivity would decrease as the deposit thickness increases due to the lack of jet penetration and convective heat transfer near the base of the deposit. The effective conductivity asymptotically approaches the conductivity of the cool air in the deposit, which decreases as the deposit insulates the target from the hot jet. In the case where the coolant temperature is lower than the metal, the growing deposit will still result in a smaller fraction of the deposit volume experiencing significant convection and a subsequent reduction in the effective conductivity, but the insulating effect of the deposit will cause the metal and thus the base of the deposit to heat up. This increase in the temperature of the stagnant air inside the deposit will increase the local conductivity, so it's reasonable to expect that the effective conductivity will asymptotically decrease as the deposit thickness increases initially. Once the heat transfer is dominated by pure conduction, the effective conductivity will increase slightly as the metal and deposit temperature increase. If future researchers attempt to assign the effective conductivity as a function of the deposit thickness, these differences in the heat transfer should be carefully considered. Using a porous model to calculate the heat transfer through the deposit should eliminate this concern because the changes in the convective and conductive heat transfer due to the deposit growth will be handled more explicitly.

While the radiative heat transfer at the relatively low metal temperatures seen in the impingement study accounted for less than 10% of the total heat transfer, at the elevated metal temperatures that occur in real engine operation the radiation is much more significant and must be accounted for in the model. An attempt to employ the surface-to-surface radiation model native to Fluent in the impingement study was made but resulted in several unresolved challenges. First, the emissivity of the solid region is applied homogenously. While this is a reasonable limitation when no deposit has accumulated on the surface, it is limiting once the deposit growth is simulated. The Hastelloy target has an emissivity of approximately 0.2 to 0.3, while the deposit has a much higher emissivity approaching that of a blackbody. While the specific conductivities of each cell in the solid region were applied simply using a user defined function, it is not clear that Fluent allows for a similar function to assign unique emissivity values.

Another challenge emerged after the mesh was morphed and the recalculation of the solution was attempted. The radiation model calculates view factors for each boundary cell and stores them to a file that is referenced whenever the solution is calculated. When the mesh is morphed, the dynamic meshing feature in Fluent often adds cells to the mesh to preserve the mesh quality. These new cells do not have a view factor to reference in the existing view factor file and the solver fails when run. The view factors must be recalculated on the new mesh, and the solution must be reinitialized and resolved. This adds significant computational time to each iteration of mesh morphing. Future efforts should find a way to extract the solution from the previous mesh and map it to the newly deformed mesh as an initial condition. This will significantly reduce the number of iterations required to reconverge the solution.

Even in the unlikely event that the deposit had an emissivity similar to the metal substrate, the surface roughness of the deposit would still cause incident radiation that is reflected to bounce of surrounding regions of the deposit, causing the deposit to absorb a higher amount of radiation. The changes in the view factors on the deposit surface must be recalculated to account for this. The ability to accurately capture this phenomenon will likely be limited by the resolution of the mesh. A very fine mesh will allow for a more robust modeling of the surface roughness of the deformed deposit and the corresponding changes in the local view factors.

www.ingramcontent.com/pod-product-compliance
Lightning Source LLC
LaVergne TN
LVHW042118190726
843493LV00006B/1519